W9-CHV-334

GORDON R. WAINWRIGHT is a well-known lecturer and is the author of *Efficiency in Reading* and *Towards Efficiency in Reading*. He is also a management skills consultant and conducts courses internationally in rapid reading and office communication skills as the director and course leader of Mandetics, his own consultancy firm.

GORDON R. WAINWRIGHT

How to Read for *Speed* and Comprehension

A SPECTRUM BOOK

PRENTICE-HALL, INC., Englewood Cliffs, New Jersey 07632

Library of Congress Cataloging in Publication Data

Wainwright, Gordon Ray.
How to read for speed and comprehension.

(A Spectrum Book)
Edition for 1972 published under title: Rapid
reading made simple.
Bibliography: p.
Includes index.
1. Rapid reading.   I. Title.
LB1050.54.W33   1977   428'.4'3   77-10468
ISBN 0-13-430769-0
ISBN 0-13430751-8 pbk.

© 1977 (revised edition) by Gordon R. Wainwright.
Published by arrangement with W. H. Allen & Co. Ltd.

Copyright 1972 by Gordon R. Wainwright.

10   9   8   7   6   5   4   3   2   1

Prentice-Hall International, Inc., *London*
Prentice-Hall of Australia Pty. Limited, *Sydney*
Prentice-Hall of Canada, Ltd., *Toronto*
Prentice-Hall of India Private Limited, *New Delhi*
Prentice-Hall of Japan, Inc., *Tokyo*
Prentice-Hall of Southeast Asia Pte. Ltd., *Singapore*
Whitehall Books Limited, *Wellington, New Zealand*

Acknowledgments

For permission to use the various extracts and passages in the exercises and tests the publishers and author gratefully acknowledge the following:

The Observer Foreign News Service for *The Troubles of Shopping in Russia, Can Man Survive?, The Man Who Built Liberia, Sri Lanka's Do-It-Yourself Food Drive; The Observer Magazine* for the six extracts entitled *It Began With the Women*; A. D. Peters and Company for *Marcuse and Revolution*; J. G. Bruton and Cambridge University Press for nine extracts from *The Story of Western Science*; Ruth Weiss for *Colourful Diversity of Culture*; the United Nations Organization for *Universal Declaration of Human Rights* (adopted by the United Nations General Assembly in 1948); Douglas Pidgeon and *The Sunday Times* for *Intelligence: A Changed View* (from an article entitled *The Expanding Mind*); Norman Moss and *The Sunday Times* for *The Think-Tank Predictors*; Edgar Dale and *The News Letter* (School of Education, Ohio State University; April 1966) for *Clear Only If Known*; Prentice-Hall, Spectrum Books for *I Could Tell by the Look in His/Her Eyes* from the book entitled *First Impressions* by Chris Kleinke; *Spatial Invasion* from the book entitled *Personal Space* by Robert Sommer; and Albert Mehrabian, Ziff-Davis Publishing Company, and *Psychology Today*, II (September 1968), 53-55, for *Communication Without Words*.

Contents

1 Initial Assessment   1

/03323

6 Final Assessment 209

7 Records and Reference 251

Preface

This book contains a training course in the skills of efficient reading. If you wish to obtain the best results from it, you will need to work through it in the same way that you would work through any other course of instruction. You will not improve your reading skills simply by reading it as you would read a light novel or a collection of short stories.

You should first of all glance through the book, reading the chapter summaries. Then you should read the introductory section which tells you precisely what will be required of you and what materials you will need. From then on you should approach each section in the order in which it appears.

At the end of the book, you will find a "Continuation" section which provides you with the means of checking your improved reading skills at intervals after completing the course. Also at the back of the book you will find suggestions for further reading on the subject of reading efficiency. Teachers will find some comments on methods of successfully tutoring reading efficiency training courses which do not require expensive equipment.

The book should prove useful to adults working in business and industry, who find that they have more reading to do at work than they can deal with effectively at present. It should also be of value to students in universities, colleges, and high schools, who are faced with extensive reading programs, and to those who are learning English as a new language. The general reader, too, should find the course helpful in becoming a faster and more efficient reader and in deriving pleasure and satisfaction from reading.

GORDON WAINWRIGHT

ONE

Initial Assessment

Before you begin the process of improving your reading skills, you should know something about the nature of the work you will be doing, the principles upon which the instruction in this book is based, and the reasons why it will be useful for you to increase your efficiency in reading. This section is designed to provide you with the preliminary information you need in order to be able to approach this course of self-instruction in such a way that you will derive the greatest possible benefit from it. At the end, your reading speed and comprehension will be tested. The results of these tests will give you your starting point for the course.

INTRODUCTION

Improved reading skills are a contribution toward general improvements in human communication. The importance of more efficient communication in business and industry, and in other fields, like education, is widely recognized. Because we communicate mainly by means of the spoken and the written word, better communication can be achieved if we can improve the skill with which we read, write, listen, and speak. This book is concerned with reading and the contribution reading improvement can make toward more effective communication.

Most of us are very slow and inefficient readers. It is generally accepted that the average reader, reading a newspaper article or similar material for a general understanding, reads at a rate of about 200 to 250 words per minute and scores between 60 and 70 percent on a comprehension test of what he has read. Evidence accumulated from reading efficiency courses over the last fifty years indicates that on material of an average level of difficulty, read for a general understanding, speeds of 300 to 800 words per minute can be achieved after a short course of training, without loss in the quality of comprehension.

Similar improvements in efficiency can be achieved on most kinds of material read for a variety of purposes. Even where increases in speed are neither possible nor desirable—for example in the detailed study of legal documents or works of literature—it is still possible for further training in the use of reading skills to enable most of us to achieve a greater degree of understanding of what we read and to eliminate time wasted in studying inessentials.

Most people are able to improve their reading skills, if they want to, but there are many for whom it is *essential* that they should be efficient readers. Managers at all levels of business and industry, administrators, teachers, technologists, public servants, and students all need to be as efficient as possible in their reading. In fact, anyone who is faced with large amounts of daily reading at work or with an extensive program of general and background reading needs to be a more efficient reader.

HOW TO APPROACH THIS COURSE

In order to make the best use of the instruction given in this book you should approach the work with appropriate attitudes of mind. *First of all, you must want to improve your reading skills.* Any learning process becomes easier if you genuinely want to take part in it. *Second, you need to have confidence in your ability to increase your reading speeds and to improve the quality of your comprehension.* If you feel that you cannot improve or are convinced that you will not improve, you are placing yourself at a disadvantage. You will still improve, regardless, but you will gain more if you have confidence in your own ability to do so. *Third, you need to be determined to follow all the instructions given you during the course.* Unless you are prepared, for instance, to practice regularly, you will not benefit as much as you might. *Fourth, to act as a spur to your efforts, you need to compete with yourself.* In a course such as this, there is nothing to be gained by competing with other students, but you will progress more rapidly if you are continually *trying* to improve upon your previous performance. This self-competition is one of your best aids to success when you are working on your own, as you are on this course (even if you are a member of a group, using this book as part of a course, you are still only concerned with your own progress). *Finally, you must work systematically as you read* and it is one of the purposes of this book to provide you with guidance in the development of systematic approaches to the reading situations you encounter in your work or in your studies.

Make these five characteristics part of your attitude toward this course and toward the process of improving your reading efficiency and your final results will be a better reflection of your abilities. A half-hearted student never achieves as much as he would if he improved his approach to his studies. In developing your approach along these lines, *it is important that you should guard against too much tenseness.* A little tension can be useful in enabling you to achieve the best possible performance each time you read. But undue strain interferes with good progress. You should try to be as relaxed and confident as possible at all times so that your mind can work free from anxieties.

Remember that *the right approach is an essential basis for the process of increasing speed in reading.* You can, in fact, increase speed by about 25 percent simply by wanting to.

Do not expect startling results immediately, however. There are courses that claim to be able to produce improvements almost overnight. You should have no such expectations. The process of permanently increasing reading speeds takes several weeks of tuition and practice, but you should be able to detect the trend toward higher speeds after two or three weeks, provided you carry out the daily practice recommended.

We should also make the point here that *this book is more concerned with enabling you to deal more effectively with reading that you* have *to do, as a part of your work or your studies, than with reading that you* want *to do, for relaxation or for general interest.* It is more concerned with situations in which a great deal has to be read in a limited amount of time than with situations in which there is plenty of time and when enjoyment and satisfaction in reading are more important than quick results.

WHAT YOU WILL NEED

As you work through this book you will require the following items:

1. A notebook for writing down in your own words points you particularly wish to remember, new words encountered for the first time, answers to comprehension questions, and so on.

2. A dictionary, so that after you have read a passage you can look up the meanings of unfamiliar words.

3. A stopwatch or a watch with a second hand for timing the reading of the passages. The standard of measurement for reading speed is "words per minute" and in order to calculate this you must know how long it takes you to read an exercise.

This is all that you will require, but *you should make sure that you have these items with you each time you work through a chapter of the book.*

In this course of self-instruction, your general aim will be to improve the **quality** and **speed** of your **comprehension in reading,** but it may be helpful if you set yourself the following rather more specific objectives:

1. To *double* your speed of reading as shown by your performance on the Initial Test passages. You may not

achieve this completely, but you should at least aim to do so.

2. To improve upon your comprehension score, as shown on the Initial Test passages, by at least 10 percent.

3. To be more flexible in dealing with *all* written materials and to make use of a wider range of reading skills.

4. To become a more critical reader.

5. To read more widely.

In the process of improving your reading skills, you will pass through three main stages. First, you will test the speed and quality of your reading comprehension as it is now. This will give you your starting point and will enable you to set yourself the targets to be reached during the course. Second, you will undergo the training, during which the *emphasis* will be on enabling you to increase the speed of your reading comprehension without loss in quality. Third, you will assess the extent of your progress at the end of the course and will be told how to maintain the improvements you have made.

During the training, you will find that there are two principal elements: **instruction** and **practice**. If you are a member of a college or company course that is using this book you will find that there is a third important element: **discussion** of the passages read for practice. During the course of the next few weeks you will be working quite hard and *the importance of regular study and practice sessions should be emphasized.* You will need to set aside about an hour a day for practice in increasing the speed and quality of your reading comprehension. You will also need to set aside an hour or so once a week to make a detailed study of each chapter in this book. Most students will, therefore, take about six weeks to work through the book.

BACKGROUND

Courses in the improvement of adult reading skills (commonly entitled "Quicker Reading," "Rapid Reading," or "Effective Reading") came into existence originally to meet the needs of business and industry. The almost complete lack of further instruction in reading once children have learned to read means that many adults are much less efficient in their reading than they should, and could, be. As with the development of any skill, there is a need for continuous instruction and guidance throughout the educational process, for skills can nearly always be improved with appropriate instruction, guidance, and practice.

Contrary to general belief, there is no virtue in slow reading for its own sake. *Slow readers are not automatically better readers.* Frequently the reverse is true and we find that slow readers are not only poor readers but that their very slowness makes the act of reading an unpleasant chore. For this reason, principally, slow readers are less likely to read as a leisure activity and the lack of wide reading in which this inevitably results can become a severe handicap in many ways. Learning how to increase speed in reading can make slow readers willing to read more and to read more widely. Thus a cycle is begun in which faster reading leads to more reading which leads to more varied reading which leads to better reading which leads to a better understanding of one's work, one's leisure interests, and oneself.

The history of reading improvement courses for adults is a surprisingly long one. Many people assume that this kind of training is a postwar phenomenon which originated in the United States. In fact, the first piece of research into eye movements in reading was carried out in France in 1878 by Emile Javal, an ophthalmologist. He discovered that the eyes did not move smoothly from left to right when a person is reading, but that they move in a series of small, jerky movements, or **"saccades,"** and that between saccades they focus or **"fixate"** on the words for about a fifth of a second. He also discovered that some readers made many fixations and read parts of the material more than once, whereas others seemed to be able to absorb a line of print in only two or three fixations and did not have to re-read the material so often. He had, in fact, discovered a fundamental difference between slow and fast readers.

The first course in reading improvement, as far as I can discover, was held at Syracuse University in New York State, in 1925. The first book on the subject, *The Art of Rapid Reading* by Walter B. Pitkin, appeared in 1929. During the thirties, more courses and books appeared and a great deal of research was carried out. During the war years, courses were provided for service personnel both in the United States and in Britain which involved programs of perception training and which helped to show that an individual's speed of perception can be increased. After the war, Harvard University Business School produced a film-aided course, which set out to train adults to read faster by having the material projected onto a screen and revealed segment by segment, much as the eyes see material in normal reading. This course is still widely used by those who think that improvements in reading can be achieved by training the eyes to make faster and more widely spaced fixations.

Since 1965, however, there has been a tendency to move away from training methods based on eye movements toward the kind of method that this book follows—that is, training that is aimed at developing in slow and inefficient readers those attitudes of mind and methods of approaching the act of reading which characterize the better, faster reader. The new approach has its basis in the realization that *poor eye movements are only a symptom and not a cause of poor reading habits.* The cure lies, therefore, not in training the eyes, but in encouraging more positive attitudes toward reading and in giving poor readers a more effective method of choice of methods for dealing with written material.

RESULTS TO BE EXPECTED

The average student who attends a reading improvement course increases his maximum speed in reading by about 80 percent, without loss of comprehension. This holds true for most types of reading material read for a variety of purposes. There is evidence that bright, young, highly motivated businesspeople tend to improve more than the average student. City people also tend to do better than country people. Some specimen results from recent courses run by the author follow, but you should remember that speeds vary with changes in purpose or material and these figures are only examples of what certain individuals achieved on material read under artificial test conditions. In practice, they may read their daily newspaper, for example, faster or slower than the figures indicate. The interesting point about the figures is that comparable material was read for similar purposes at both the beginnings and ends of the courses and it is, therefore, the percentage increases in reading speeds which you should notice.

**RESULTS OF SOME INDIVIDUAL STUDENTS
WHO HAVE ATTENDED THE AUTHOR'S COURSES**

Student	*Start of Course*		*End of Course*		*% Increase Speed*
	Speed w.p.m.	*Comprehension %*	*Speed w.p.m.*	*Comprehension %*	
1	212	67	432	77	+103
2	125	63	273	70	+118
3	160	67	240	ᴜ0	+75
4	154	50	273	83	+77
5	234	93	339	93	+45
6	233	80	467	87	+100
7	166	93	348	83	+110
8	220	80	380	70	+73
9	240	60	420	60	+75
10	350	67	571	70	+63
11	266	50	666	60	+150
12	149	87	475	97	+218
13	253	70	586	73	+131
14	211	83	365	77	+73
15	389	83	1104	97	+184
16	277	93	466	80	+68
17	137	83	176	80	+29
18	153	77	435	90	+185
19	215	80	585	90	+170
20	246	80	500	83	+103
21	318	83	510	80	+60
22	218	77	287	87	+32
23	167	67	233	73	+40
24	180	80	241	70	+33
25	146	87	368	70	+152
26	168	83	262	80	+57
27	194	70	250	67	+28
28	188	67	220	77	+17
29	171	80	277	77	+59
30	141	67	230	70	+61
31	160	77	212	70	+33

w.p.m. = words per minute

RESULTS OF SOME INDIVIDUAL STUDENTS (*cont.*)

| Student | Start of Course | | End of Course | | % Increase Speed |
	Speed w.p.m.	Comprehension %	Speed w.p.m.	Comprehension %	
32	180	93	274	87	+50
33	188	83	285	83	+52
34	261	73	530	73	+103
35	336	77	502	80	+50
36	344	73	1000	67	+190
37	201	87	306	63	+52
38	250	77	360	63	+44
39	300	70	409	77	+36
40	193	73	300	77	+55
41	175	70	283	83	+62
42	282	83	668	73	+140
43	287	80	1050	93	+265
44	206	80	266	90	+22
45	263	70	533	83	+101
46	394	83	750	73	+90
47	266	97	376	90	+41
48	160	77	225	77	+41
49	169	67	231	73	+37
50	230	73	384	80	+67
51	231	70	277	83	+20
52	259	70	350	67	+35
53	256	100	513	100	+100
54	311	97	744	87	+179
55	230	90	558	90	+143
56	240	83	458	90	+90
57	218	90	432	87	+99
58	177	83	498	83	+180
59	204	83	561	97	+190
60	225	70	624	77	+178

w.p.m. = words per minute

ASSESSING QUALITY OF COMPREHENSION

It is fairly easy to assess objectively the *speed* of a person's reading comprehension by using the single reading of a set of passages as a basis, measuring how long the reading takes and expressing this as "words per minute." But it is very difficult, if not impossible, with all but the simplest material to assess the *quality* of a person's reading comprehension with any degree of reliability. Quality of comprehension can fluctuate, or appear to fluctuate, for many reasons. The reader may allow his attention to wander momentarily at a key point in the material, the writer may express himself inadequately, people may be talking in the same room where a person is reading, the reader may be confused by an unfamiliar word. These, and many other factors, may cause the quality of reading comprehension to suffer temporarily. In working through this book, you should remember that your scores *will* fluctuate and that it is quite natural that they should do so.

As you will see, most of the tests in this book set out to assess your ability to **recall** information immediately after having read it (Section A—Retention) and your ability to understand the significance or **interpret** the meaning of parts or the whole of the passage (Section B—Interpretation). Most of the tests, and especially the initial and final tests, are *objective*, in the sense that you have to select the most suitable answer from four alternatives, but on some passages *open* questions have been set. Your answers to these should be as brief as possible and suggested answers are included in the Answers to Comprehension Tests section on pages 257-262. There are other ways of testing the quality of comprehension, and in your daily practice reading of newspapers, journals, books, and so on, you may like to try two of them as useful alternatives to the methods used here.

The first method is **summary**. After you have read an article or a chapter in a book, try briefly to summarize in your own words the essence of what you have read. The other method is **discussion**. Whenever you can, discuss your reading with others. These methods will also help you to *improve* the quality of your comprehension—a point that is discussed more fully in Chapter 2.

You will, in fact, find that some passages in this book do require you to attempt a summary. If you are a member of a class, you will also be able to discuss some of the questions in the Discussion section of the tests as well as other questions of your own arising out of the content of the passages. So you will gain a little practice in using these methods during this course. The main comprehension test, however, remains one of retention and of interpretation and it is the results for Section A and Section B that you will record on the Progress Graph.

PROCEDURE FOR THE READING EXERCISES

So that you will be familiar with the procedure to be followed in reading the practice passages during the course, there follows a short trial exercise for you to attempt. The **procedure** for reading this and other passages—except where you are given different instructions—is as follows:

1. Have your stopwatch, or a watch with a second hand, ready and at a convenient point begin timing and begin reading.
2. Read the passage through **once only** as quickly as you can without loss of comprehension and note the time taken.
3. Answer Sections A and B of the Comprehension Test.
4. Convert the time taken to read the passage into "words per minute," using the Reading Speed Conversion Table on page 254.
5. Check your answers to Sections A and B of the Comprehension Test against those given on pages 257-262.
6. Enter the results on the Progress Graphs on page 256.
7. Deal with Section C of the Comprehension Test.
8. Turn to the next passage and follow the same procedure. *This last point does not apply to this trial exercise, of course.*

Follow this procedure for *every* exercise, unless otherwise instructed. *Read the instructions again, carefully, if you are at all unsure on any point.*

A NOTE ON THE HISTORY OF SCIENCE PASSAGES

The passages on Seventeenth, Nineteenth and Eighteenth Century Science (used for the Initial, Final, and Three-Month Follow-Up Tests) are taken from *The Story of Western Science* by J. G. Bruton (Cambridge University Press). The vocabulary has been brought within the limits of West's *General Service List of English Words* as far as possible, and the sentence and paragraph structure are simplified. Tests for balancing have been conducted by the author to ensure that these passages are comparable in their level of difficulty. This means that the important measurements of reading speed in this course are conducted on similar material (performance on the passages on the Russian Revolution may also be compared one with another). Any improvements indicated will thus be "real" improvements in reading performance.

TRIAL READING EXERCISE

Proceed now to the trial exercise. Read it through **once only** as quickly as you can without suffering loss of comprehension. You are, after all, trying to find out how quickly you can read at the moment.
Set your watch and begin reading NOW.

THE TROUBLES OF SHOPPING IN RUSSIA
by Dev Murarka

A large crowd gathered outside a photographic studio in Arbat Street, one of the busiest shopping streets in Moscow, recently. There was no policeman within sight and the crowd was blocking the pavement. The centre of the attraction—and amusement—was a fairly well-dressed man, perhaps some official, who was waving his arm out of the ventilation window of the studio and begging to be allowed out. The woman in charge of the studio was standing outside and arguing with him. The man had apparently arrived just when the studio was about to close for lunch and insisted upon taking delivery of some prints which had been promised to him. He refused to wait so the staff had locked the shop and gone away for lunch. The incident was an extreme example of a common attitude in service industries in the Soviet Union generally, and especially in Moscow. Shop assistants do not consider the customer as a valuable client but as a nuisance of some kind who has to be treated with little ceremony and without much concern for his requirements.

For nearly a decade, the Soviet authorities have been trying to improve the service facilities. More shops are being opened, more restaurants are being established and the press frequently runs campaigns urging better service in shops and places of entertainment. It is all to no avail. The main reason for this is shortage of staff. Young people are more and more reluctant to make a career in shops, restaurants and other such establishments. Older staff are gradually retiring and this leaves a big gap. It is not at all unusual to see part of a restaurant or a shop roped off because there is nobody available to serve. Sometimes, establishments have been known to be closed for several days because of this.

One reason for the unpopularity of jobs in the service industries is their low prestige. Soviet papers and journals have reported that people generally consider most shop assistants to be dishonest and this conviction remains unshakeable. Several directors of business establishments, for instance, who are loudest in complaining about shortage of labour, are also equally vehement that they will not let their children have anything to do with trade.

The greatest irritant for the people is not the shortage of goods but the time consumed in hunting for them and queueing up to buy them. This naturally causes ill-feeling between the shoppers and the assistants behind the counters, though often it may not be the fault of the assistants at all. This, too, damages hopes of attracting new recruits. Many educated youngsters would be ashamed to have to behave in such a negative way.

Rules and regulations laid down by shop managers often have little regard for logic or convenience. An irate Soviet journalist recently told of his experiences when trying to have an electric shaver repaired. Outside a repair shop he saw a notice: "Repairs done within 45 minutes." After queueing for 45 minutes he was asked what brand of shaver he owned. He identified it and was told that the shop only mended shavers made in a particular factory and he would have to go to another shop, four miles away. When he complained, the red-faced girl behind the counter could only tell him miserably that those were her instructions.

All organisations connected with youth, particularly the Young Communist League (Komsomol), have been instructed to help in the campaign for better recruitment to service industries. The Komsomol provides a nicely-printed application form which is given to anyone asking for a job. But one district head of a distribution organisation claimed that in the last 10 years only one person had come to him with this form. "We do not need fancy paper. We need people!" he said. More and more people are arguing that the only way to solve the problem is to introduce mechanisation. In grocery stores, for instance, the work load could be made easier with mechanical devices to move sacks and heavy packages.

The shortages of workers are bringing unfortunate consequences in other areas. Minor rackets flourish. Only a few days ago, *Pravda*, the Communist Party newspaper, carried a long humorous feature about a plumber who earns a lot of extra money on the side and gets gloriously drunk every night. He is nominally in charge of looking after 300 apartments and is paid for it. But whenever he has a repair job to do, he manages to screw some more money from the flat dwellers, pretending that spare parts are required. Complaints against him have no effect because the housing board responsible is afraid that they will be unable to get a replacement. In a few years' time, things could be even worse if the supply of recruits to these jobs dries up altogether.

800 words

Write down the time taken to read this passage and then attempt the Comprehension Test.

COMPREHENSION TEST

Select the most suitable answer in each case.
Do **not** *refer back to the passage.*

A. Retention

1. The large crowd in Arbat Street was gathered outside:
    a) a restaurant.
    b) a shop.
    c) a block of apartments.
    d) a photographic studio.

2. Shop assistants consider the customer as:
    a) a valuable client.
    b) an equal.
    c) a nuisance.
    d) an enemy.

3. One reason given in the passage for the unpopularity of jobs in service industries is:
    a) long hours.
    b) low prestige
    c) low wages.
    d) the hard work.

4. More and more people are arguing that the only way to solve the problem is to:
    a) rope off parts of restaurants.
    b) introduce mechanization.
    c) offer high wages.
    d) have a campaign for better recruitment.

5. The man who earned "a lot of extra money on the side" was:
    a) a plumber.
    b) a journalist.
    c) a repairer of electric shavers.
    d) an official.

B. Interpretation

6. In trying to improve service facilities, the Soviet authorities are having:
   a) great success.
   b) some success.
   c) little success.
   d) no success.

7. The man who had been locked in the photographic studio was:
   a) angry.
   b) unconcerned.
   c) embarrassed.
   d) afraid.

8. The rules and regulations laid down by shop managers are generally:
   a) helpful.
   b) necessary.
   c) impossible to carry out.
   d) unhelpful.

9. The girl in the shaver repair shop was "red-faced" because she was:
   a) healthy.
   b) embarrassed.
   c) angry.
   d) crying.

10. One effect of the shortage of workers in service industries is that:
    a) corruption is encouraged.
    b) prices rise.
    c) wages rise.
    d) mechanization has been introduced.

Convert the time taken to read the passage into "words per minute" by using the Reading Speed Conversion Table on page 254. Enter the result on the Progress Graph on page 256.

Check your answers to the Comprehension Test against the answers given on page 257. Enter the result on the Progress Graph on page 256.

C. Discussion—discuss one of these questions (orally if in a group, in writing if studying alone). *You may refer back to the passage.*

11. What are the similarities and differences between the Russian and American attitudes to service industries and to the workers in those industries?

12. Do you feel that the passage presents an accurate picture of this aspect of life in Russia today? Give your reasons.

13. What do you think the Russians could do to improve service facilities generally?

14. What reasons can you find that are not mentioned in the passage for the unpopularity of jobs in service industries in Russia?

15. How would you apportion responsibility for the poor service facilities in Russia between the government and the people themselves?

INITIAL TESTS

The tests that follow will give you a more accurate assessment of your present speed and quality of reading comprehension than you obtained on the trial exercise.

Altogether, these tests will take you approximately one hour to complete. Select a time and place that will enable you to attempt them without interruption. Remember that you will be asked to select a similar time and place for the Final Tests, so that valid comparisons can be made and improvements assessed with a reasonable degree of reliability.

At this point, spend a few seconds (about fifteen) looking through the tests, noting the subject matter and the length of each and the location of the test questions. Quickly glance through the Initial Tests on pages 18 to 29 and do not try to read anything in detail. All you require at this stage is an idea of the size of the task that faces you. The tests are not very difficult, so try to avoid regarding this as an "examination." Look upon it as an interesting and informative exercise. Even if you are a member of a group using this book as part of a course, *you are not competing with anyone else and your scores are entirely your own affair.*

Make sure that you have with you your watch, your note-book, and a pen or pencil. Preferably, sit at a desk or table and place these articles within easy reach.

You are now ready to begin. When you have completed the tests, you will be given further instructions.

Begin NOW.

Instructions

Read through the passage once only *as quickly as you can without loss of comprehension.*
Begin timing and start reading NOW.

SEVENTEENTH CENTURY SCIENCE—I

by J. G. Bruton

Historians of science are divided in their explanation of the change in western man's outlook which occurred in the seventeenth century. Some consider seventeenth century science as a sudden advance in a process which had been going on since the earliest days of history. This view seems reasonable when we remember how Copernicus was influenced by Ptolemy, Galileo by Aristotle and Archimedes, and how much Vesalius and Harvey owed to Galen. Others believe that the growth of modern science was the result of economic changes. The growth of trade and the beginnings of capitalism can be clearly seen in the Middle Ages: but in the sixteenth and seventeenth centuries there was also an increase in trade and industry which took place at the same time that science was developing. Science developed in Italy at a time when the country was rich. As trade and industry moved towards the Atlantic, the centre of scientific advance also moved to Holland, France and England.

Early in the sixteenth century, prices began to rise all over Europe and continued to rise until the middle of the seventeenth century. Then began a period of international trading and competition, and during this period scientific societies were formed in England and France. Merchants saw these societies as a means of obtaining inventions which would make it possible to reduce costs and to increase production.

There are many examples of the way in which technology and science influence one another. The medieval clock was the origin of the work on the pendulum done by Galileo. The needs of navigation led to the interest in astronomy which produced the work of Copernicus, Tycho Brahe, Kepler, Galileo and Newton. The manufacture of spectacles, especially in Holland, made possible the invention of the telescope and the microscope. The needs of the mining industry in Germany, Hungary and Holland led to experiments on the vacuum, to the invention of the air-pump by Otto von Guericke and to Boyle's work on gases. The idea of the pump was the basis of Harvey's researches on the circulation of the blood and of Newcomen's invention of the steam-engine in 1712.

Familiarity with machines of all kinds made people ready to accept the idea of a mechanistic universe which was so different from the medieval ideal of the universe. Another characteristic of the new movement was the attention it paid to exact measurement. This led to the invention of Galileo's thermometer, Torricelli's barometer, Gunter's slide rule, Gascoigne's micrometer and Huygen's pendulum clock.

The most important characteristic of the new scientific outlook was the use of experiment. The Greek philosophers and the scholars of the Middle Ages did perform some experiments, but it was not their general practice. They worked almost entirely with their heads and scarcely at all with their hands. The "new philosophy" of science was the result of a combination of this tradition and that of the manual worker.

490 words

Write down the time taken to read this passage and then attempt the Comprehension Test.

COMPREHENSION TEST

Select the most suitable answer in each case.
Do not *refer back to the passage.*

A. Retention

1. The writer states that some historians believe the growth of modern science was the result of:
   a) pure chance.
   b) political developments.
   c) economic changes.
   d) geographical factors.

2. Science developed in Italy at a time when the country was:
   a) industrially advanced.
   b) rich.
   c) poor.
   d) backward.

3. Merchants saw the scientific societies, which were formed in England and France, as:
   a) a means of obtaining inventions that would reduce costs.
   b) a threat to their own existence.
   c) a means of avoiding work.
   d) a means of changing the structure of society.

4. The medieval clock was the origin of the work done on the pendulum by:
   a) Harvey.
   b) Aristotle.
   c) Newton.
   d) Galileo.

5. Familiarity with machines of all kinds made people:
   a) contemptuous of them.
   b) ready to accept the idea of a mechanistic universe.
   c) afraid of them.
   d) ready to believe anything that the scientists told them.

B. Interpretation

6. The development of trade and industry and the advances in science in the seventeenth century:
   a) had no effect on each other.
   b) were planned by the scientists.
   c) were closely linked.
   d) were planned by politicians.

7. The period of international trading and competition which began in the middle of the seventeenth century:
   a) caused prices to rise.
   b) halted rising prices.
   c) had no effect upon prices.
   d) restricted the growth of science.

8. Seventeenth-century scientists were primarily interested in:
   a) the practical applications of science.
   b) the relationship between science and the arts.
   c) making money from inventions.
   d) finding a scientific basis for the existence of God.

9. Seventeenth-century scientists thought exact measurement was important:
   a) because it emphasized the difference between science and art.
   b) because it made science a respectable subject of study.
   c) because it meant they could use instruments they had invented.
   d) because it helped to provide a firm basis for their conclusions.

10. The use of experimentation was the most important characteristic of the new scientific outlook because:
    a) it helped scientists to test the truth of their ideas.

b) experiments had never previously been carried out by scientists.

c) it emphasized the difference between science and the arts.

d) it made scientists' work understandable to the layperson.

Convert the time taken to read the passage into "words per minute" by using the Reading Speed Conversion Table on page 254. Enter the result on the Progress Graph on page 256.

Check your answers to the Comprehension Test against the answers given on page 257. Enter the result on the Progress Graph on page 256.

Proceed to the next reading exercise.

Instructions

Read through the passage once only *as quickly as you can without loss of comprehension.*

Begin timing and reading NOW.

SEVENTEENTH CENTURY SCIENCE—II

by J. G. Bruton

By the end of the sixteenth century, there existed a large class of manual workers—miners, glass workers, navigators, makers of instruments. Many of these had no education, but, because of economic competition, they experimented with their materials and made improvements and small discoveries. The rise of a capitalist society brought about a meeting of the workers and the upper classes who began to take an increasing interest in technology.

The Marxist explanation of the establishment of science in terms of economics suffers from one important weakness. Galileo and Newton both worked in universities and had little economic motive. Newton especially was greatly interested in the connection between science and religion; the same is true of Boyle. Harvey and the earlier microscopists were similarly not interested in business. The society of the seventeenth century offered conditions under which science was able to develop, but these conditions were not directly responsible for the development of modern science.

Two men who had a great influence on thought in the seventeenth century were Francis Bacon (1561-1626) and Rene Descartes (1596-1650); the first spread the idea of the experimental method and of the control of nature, the second popularised the idea of a mathematical and mechanistic universe.

Bacon lived during the period when England was rapidly developing industrially. He believed from a very early age that science could change man's life by controlling nature, and that the secret lay in applied science. He believed in the value of experiment and in the collection by observation and experiment of facts from which it would be possible to form generalisations. This inductive method, based on observation, was very different from the method of Aristotle, which was deductive, based on thought. Bacon, however, had no real knowledge of scientific method and it is strange that he did not recognise the importance of the work of such people as Harvey, who was his own doctor, or of William Gilbert and Galileo.

The simplest form of induction is mere counting, the collection of examples. Bacon did not accept this type of induction. Most of us would accept the generalisation that all swans are white, on the evidence of a large number of examples, but this is not true, since black swans exist in Australia. Bacon used instead a system of induction by elimination, by which, instead of collecting only positive examples, he collected also negative examples which his theory eliminated. His method was different from normal scientific method. He did not see the great importance of hypothesis in deciding which facts to look for and which experiments to perform.

Bacon himself made no important discoveries but his influence was felt by the founders of the Royal Society and by the French encyclopedists of the eighteenth century. Macaulay said that he moved the minds which moved the world.

Descartes's method was different from Bacon's, for it was mathematical and deductive. Beginning from very simple principles, Descartes built a picture of the universe which had an enormous effect on the thought of his time, just when the influence of Aristotle was beginning to lose power.

520 words

Write down the time taken to read this passage and then attempt the Comprehension Test.

COMPREHENSION TEST

Select the most suitable answer in each case.

Do not *refer back to the passage.*

A. Retention

1. Many manual workers experimented with their materials and made improvements because of:
    a) pressure from their employers.
    b) economic competition.
    c) a desire to imitate the scientists.
    d) natural curiosity.

2. The idea of a mathematical and mechanistic universe was popularized by:
    a) Descartes.
    b) Newton.
    c) Bacon.
    d) Galileo.

3. Bacon lived during the period when England was:
    a) more interested in Shakespeare than science.
    b) economically depressed.
    c) rapidly developing industrially.
    d) at war with Germany.

4. Black swans exist in:
   a) Australia.
   b) France.
   c) Greece.
   d) Germany.

5. Bacon himself:
   a) made many important discoveries.
   b) made few important discoveries.
   c) made some important discoveries.
   d) made no important discoveries.

B. Interpretation

6. The rise of a capitalist society:
   a) was hindered by scientific advances.
   b) was helped by scientific advances.
   c) was unaffected by scientific advances.
   d) was controlled by scientific advances.

7. The Marxist explanation of the establishment of science in terms of economics suffers from one important weakness because:
   a) Marx was not a scientist.
   b) economic factors do not affect the progress of science.
   c) many scientists were not familiar with Marx's ideas.
   d) many scientists had little or no economic motive in their work.

8. The inductive method is the method by which:
   a) general laws are made by studying particular facts or examples.
   b) scientists arrive at their conclusions.
   c) particular facts or examples are explained by general laws.
   d) the laws of nature are explained by scientists.

9. The inductive method was based on observation because:
   a) scientists had never done this before.
   b) one or two examples had to be collected.
   c) this distinguished it from the deductive method.
   d) a large number of examples had to be collected.

10. Francis Bacon had a great influence on thought in the seventeenth century:
    a) because of his ideas on experimental method.
    b) because he made many important discoveries.
    c) because he developed the deductive method.
    d) because he popularized the idea of a mechanistic universe.

Convert the time taken to read the passage into "words per minute" by using the Reading Speed Conversion Table on page 254. Enter the results on the Progress Graph on page 256.

Check your answers to the Comprehension Test against the answers given on page 257. Enter the results on the Progress Graph on page 256.

Proceed to the next reading exercise.

Instructions

Read through the passage **once only** *as quickly as you can without loss of comprehension.*

Begin timing and begin reading NOW.

SEVENTEENTH CENTURY SCIENCE—III

by J. G. Bruton

Descartes based his philosophy on mathematics, which has the two great qualities of clarity and certainty. He was himself a distinguished mathematician and the founder of co-ordinate geometry, which is a combination of algebra and geometry.

The science and philosophy of Descartes were presented like geometry. His starting-point was the axiom: "I think, therefore I exist"; from this he deduced that God existed and that matter existed. He believed that we have in our minds inborn ideas which have become confused by mistakes in what we have read, and he insisted on the necessity for accepting only those ideas which are clear and reasonable.

He expressed his ideas very simply, with the result that his works were very popular. But, like Bacon, he did not value or understand the work of Galileo and did not recognise the full importance of experiment in establishing the basic principles of science. Although he was not a great scientist, Descartes was the first modern philosopher. He believed that mind and matter were independent of one another, and that matter was self-operating like a machine, and in this class he placed animals. He saw no conflict between religion and science since they were concerned with different matters.

Descartes did a lot of his thinking in bed in the morning or when he was sitting in a really warm room. At the age of 55 he was invited to Sweden by Queen Christine to be her tutor; she sent a warship to fetch him. She liked to begin her studies at 5 a.m.; Descartes as a result caught a cold from which he died.

The study of science largely outside the universities led to the formation of scientific societies. One of the earliest societies or academies for the study of science was founded in Rome in 1601; Galileo was one of its members. Another society was established in Florence by the Medici family; one of its chief interests was the perfecting of scientific instruments, about which it published a book which became very popular in the eighteenth century.

Similar groups were formed in France, the most famous being the Academie des Sciences, established in Paris in 1666; it consisted of 20 members, each of whom received a pension from the King. It grew in size and importance and counted among its members all the most famous people interested in science.

In England the Royal Society was founded in 1662 by Charles II. Its object was "the improving of Natural Knowledge (knowledge of nature) by experiment." The centre of the Society was Gresham College, founded by Sir Thomas Gresham for the teaching of such subjects as Geometry, Astronomy, Geography and Navigation. It became the centre of scientific activity, more important than the universities. Merchants took great interest in the Society, which set up committees to investigate the history and the technical needs of such trades as shipping, mining, brewing, the manufacture of wool and the making of bread.

490 words

Write down the time taken to read this passage and then attempt the Comprehension Test.

COMPREHENSION TEST

Select the most suitable answer in each case. ·
Do **not** *refer back to the passage.*

A. Retention

1. Descartes based his philosophy on.
   a) astronomy.
   b) geography.
   c) mathematics.
   d) biology.

2. Descartes' starting point was:
   a) "I think I exist."
   b) "I exist, therefore I think."
   c) "I exist, I think."
   d) "I think, therefore I exist."

3. Although he was not a great scientist, Descartes was:
   a) the first modern philosopher.
   b) the first mathematician.
   c) the greatest philosopher.
   d) the first modern novelist.

4. Queen Christine liked to begin her studies at:
   a) 3 a.m.
   b) 4 a.m.
   c) 5 a.m.
   d) 6 a.m.

5. The object of the Royal Society was:
   a) "the helping of Industry by Scientific Research."
   b) "the improving of Natural Knowledge by experiment."
   c) "to become the Centre of Scientific Activity."
   d) "to advise the King about the Progress of Science."

B. Interpretation

6. Mathematics has the two great qualities of clarity and certainty:
   a) because it follows strict and logical rules.
   b) because it is the basis of all scientific activity.
   c) because it has to produce definite answers.
   d) because it is the language of science.

7. Descartes did not recognize the full importance of experiment in establishing the basic principles of science because:
   a) his approach was inductive.
   b) he did not understand the work of Bacon.
   c) his approach was deductive.
   d) his approach was intuitive.

8. Descartes regarded animals as machines because:
   a) he did not believe they could think.
   b) they were used by men as machines.
   c) he did not like animals.
   d) they were inferior to men.

9. Scientific societies were formed in Europe in order to:
   a) enable scientists to keep their findings to themselves.
   b) secure financial support for scientific activity.
   c) publish books about science.
   d) encourage the progress of science.

10. Merchants took great interest in the work of the committees set up by the Royal Society because:
    a) they did not trust the scientists.
    b) they hoped the work would produce results they could profit from.
    c) they had given the Society large sums of money.
    d) they hoped to please the King.

Convert the time taken to read the passage into "words per minute" by using the Reading Speed Conversion Table on page 254. Enter the result on the Progress Graph on page 256.

Check your answers to the Comprehension Test against the answers given on page 257. Enter the result on the Progress Graph on page 256.

When you have completed the Initial Tests, calculate your *average* score by adding the three results together and dividing by three. Do this for both reading speed and comprehension. *Enter the results on the Progress Graphs on page 256 and regard this as your initial score and starting-point for the course.*

A QUESTIONNAIRE

Now, answer the following short questionnaire, which will enable you to assess other needs from this course of training. Answer as honestly as you can. **Answer in writing.**

1. Do you really *want* to improve your reading skills?
2. Do you feel that you *can* improve your reading skills?
3. Are you prepared to practice improving your reading skills for about an hour a day for the next six weeks or so?
4. Do you regard yourself as generally a slow reader or a fast reader?
5. Do you regard yourself as generally a good reader or a poor reader?
6. Do you generally remember much or little of, for example, a newspaper article after a single reading?
7. As you read, do you often go back to read parts of the material again?
8. As you read silently, are you aware of mouthing each word or "hearing" the words in your mind?

9. Fix your eyes briefly on the word in italics in the sentence below. Without moving your eyes, how many words can you read to the left and the right of the italicized word?

   Do not move your eyes *while* you look at the word.

10. How good is your concentration when you are reading? Can you, for example, become completely absorbed in your reading when people are talking or when there are other distractions? Can you read for long without your attention wandering?

11. How often do you come across words whose meanings you do not clearly understand?

12. As you are reading, can you anticipate the general sense of what follows?

13. Are you always sure about your purpose in reading every piece of written material you encounter?

14. How much general reading do you do during your leisure time?

15. Are you interested in a wide range of subjects and activities apart from those directly connected with your work or any hobby you may have?

16. Can you skim quickly through written material to obtain a general idea of the contents?

17. Do you feel relaxed or rather tense when you have a great deal of reading to do in a relatively short time?

18. How critical is your approach to reading? Do you continually try to analyze and evaluate what you read? Do you think you *need* to read critically?

19. Do you quickly adapt your speed of reading to changes in the level of difficulty of the material?

20. Do you feel you have a clear understanding of the principal elements of grammatical structure in English and of how writers construct their material?

ANSWERS TO THE QUESTIONNAIRE

Compare your answers with the comments given below on the answers to these questions.

1. Your answer must be an emphatic *Yes* if you are to derive the maximum benefit from this course of training.

2. You must convince yourself, if you are not already convinced, that you *can* improve your reading skills. After all, if you were already a perfectly efficient reader, you would not have decided to work through this book.

3. If you are not prepared to *practice* for about an hour a day over the next six weeks or so, you will not gain as much as you might from this course and any gains you do make are likely to be only temporary. *Regular daily practice is essential if you are to achieve permanent improvements.*

4. Most students will regard themselves as slow readers, but it should be pointed out that even fast readers have, in the past, benefited greatly from courses like this one.

5. Most students will, if they are honest with themselves, admit that they are poor readers, though even good readers can become better readers.

6. Most readers quickly forget many of the details of what they have read, after a single reading. This course will help you to improve your skill in retaining information.

7. If your answer was "Yes," you should pay particular attention to the section on Avoiding Regressions.

8. Many teachers of reading improvement regard sub-vocalization, or "hearing" the words in one's mind as one reads silently, as a fault. My advice to you is to disregard this. There is no evidence that subvocalization prevents reading speeds from being increased.

9. The average slow reader will be able to see one word reasonably clearly on either side of the word italicized; the average fast reader will be able to see two words on either side.

10. Your ability to concentrate should improve during this course.

11. A method for widening vocabulary is suggested in Chapter 5.

12. The slow reader, because of his slowness, has difficulty in using his anticipation skills to the full in reading. *Faster reading will bring better anticipation.*

13. An inadequate sense of purpose in reading is one of the main reasons for ineffectual reading. The importance of clear definition of purpose is dealt with in Chapter 3.

14. You will be reminded during the course about the value of general reading in helping to improve reading skills.

15. There is evidence to suggest that people with wide interests are generally better readers.

16. Slow readers are invariably incapable of skimming written material effectively for a general understanding or for specific information, and at the beginning of a course of training, many refuse to believe that it is possible to gain anything at all from skimming. Chapter 4 deals with skimming.

17. Greater confidence, resulting from improved reading skills, will help you to avoid tension when reading "against the clock."

18. Everyone needs to read critically. The importance of using your critical faculties when reading is discussed in Chapter 3.

19. The importance of a flexible approach to reading and in factors to be considered in achieving this are discussed in Chapter 4.

20. It is beyond the scope of this book to provide readers with an understanding of the principal elements of grammatical structure in English, but certain fundamental points are discussed in the section "How Written Materials are Constructed" (Chapter 3).

If your answers to any of these questions indicate that there is room for improvement, you should pay particular attention to the chapters in which these points are considered. You might like to note these chapters in your notebook so that you can be sure to give them special study.

PRACTICE FOR THE COMING WEEK

Make a list of everything you read this week, together with times taken to read various materials.

Summarize at least one article from a newspaper or magazine every day.

Every time you read, make a conscious effort to read a little faster than you find comfortable.

CHAPTER SUMMARY

1. **Introduction**
   a) Most people are slow, inefficient readers.
   b) Most people can increase their reading speed, without loss of comprehension, by an average of 80 percent.
   c) There are some people who *must* be able to read well.

2. **How to Approach this Course**
   a) You must *want* to improve your reading skills.
   b) You must feel you *can* improve your reading skills.
   c) You must follow the course contained in this book conscientiously and with determination to succeed.
   d) You must compete *with yourself* and continually try to improve upon your previous performance.
   e) You must develop systematic approaches toward reading.
   f) You must guard against too much tenseness in your approach.
   g) This book is more concerned with helping you to improve your reading of materials you **have** to deal with than with those you **want** to read, though these will benefit, too.

3. **What You Will Need**
   a) A notebook.
   b) A dictionary.
   c) A stopwatch or a watch with a second hand.
   d) One of your objectives will be to *double* your reading speed.
   e) Another will be to improve your comprehension score by at least 10 percent.

4. **Assessing Quality of Comprehension**
   a) Comprehension scores will fluctuate and it is quite natural that they should do so.
   b) Two methods of testing comprehension which you should try in your daily reading are:
      i) *summary*—try to express the essence of what you have read, in your own words.
      ii) *discussion*—whenever you can, discuss your reading with others.

TWO

Reading and Readers

THE NATURE OF THE READING PROCESS

To understand the nature of the basic differences between the
techniques used by slower readers and the more effective
techniques used by efficient readers, we must first understand
the nature of the reading process. If you stand at a window
overlooking a busy road and watch a car pass you from left to
right, your eyes will appear to move smoothly because they are
focused on the car. If, however, you wait until there is no traffic
and try to follow an imaginary car as it moves from left to right
anyone who watches your eyes while you do this will tell you
that they move in a series of small jerks. This is what happens
when you read. As your eyes move from left to right along a
line of print, they make a series of small jerky movements
("saccades"), stopping momentarily on each word or group of
words. These pauses are called "fixations" and each one lasts for
about a third of a second. This is when the actual reading is done,
because, it is now believed, there is a mechanism in the brain

which "switches off" vision while the eyes are moving from one fixation point to the next to prevent things from always appearing blurred. The eye can thus only accept information in detail when it is allowed to fixate momentarily on the material before it.

The slow reader finds that he has to fixate on every word in order to understand what he reads. The efficient reader, on the other hand, either naturally or through training, has learned to widen his **eye span** (the amount he can see clearly at each fixation) and to see written material more in terms of groups of words than as single words, each of which is equally independent and important. There are many films and mechanical devices available that claim to be able to help any reader to widen his eye span, but no one has yet been able to produce evidence that they are any more effective than simply trying to read faster. In fact, because you usually cannot read faster without widening your eye span, if your reading speeds increase as a result of trying to read faster, you have widened your eye span without necessarily realizing it. The one may be said to follow naturally from the other.

But **speed** is only one aspect of reading that we are seeking to improve, though it is the one with which we are primarily concerned in this book. **Quality** must also be improved. Or at least it must not be allowed to suffer in the quest for greater speed. In discovering how to improve both the speed and the quality of comprehension in reading, we need to take a closer look at the nature of comprehension itself and attempt to assess some of the factors that affect it.

THE NATURE OF COMPREHENSION

Clearly, there is a strong subjective, personal element in comprehension and it is frequently possible for there to be several "correct" understandings of any particular piece of reading matter. Among the many intellectual abilities required for comprehension, the following are probably the most important:

The ability to retain information and recall it when required.
You should be able to remember a reasonable proportion of
the factual information or the ideas expressed in a passage if
you are to be able to say you have understood it. Retention
immediately after having read a passage will be higher than
retention some time later, so that the Retention section of the
Comprehension Tests will help you to assess how much you are
able to recall before you really begin to forget the material.
Rates of forgetting vary from one person to another. More is
forgotten in the first few minutes after having read something
than at any other time. Some readers will need to pay more
attention to improving this aspect of comprehension than others,
depending on their needs from reading. Results on the Retention
sections should be used to see how much improvement in
retention is necessary. These tests will also help to bring about
improvements. Occasionally, you may wish to allow some
time to elapse between reading a passage and answering the
comprehension test to find out how much you can remember
of what you have read after the initial post-reading "forgetting
period."

The ability to select important points. If you have understood
what you have read, you should be able to differentiate major facts
and ideas from minor ones. There is an element of subjectivity
here in that what one reader will regard as important another
will not, but it is usually possible, through discussion, to arrive
at some form of general agreement on the important points in
any piece of material. If, therefore, you are in doubt about the
main points, discuss your assessment with others who have read
the material. Part of this process involves being able to select
relevant information and discard the irrelevant. Before you assess
which of the points a writer makes are the important ones, you
should discard any that are irrelevant to his theme or purpose.
In other words you should beware of imposing your own views
on the writer.

The ability to interpret information and ideas. This follows from
the previous point. In addition to being able to select the
important and the relevant, you should be able to understand the
meaning and significance of these facts or ideas. This is necessary
if you are to make the maximum use of what you learn from

your reading, and it is part of the process of being able to justify your assessment to others.

The ability to make deductions from what has been read. Often points are not stated explicitly and it is reasonable to expect a reader who has understood the material to be able to deduce certain things for himself from information given. Similarly, it is reasonable to expect a reader to be able to draw inferences and be aware of implications in the material. It may not be possible for you to arrive at these by the same clear processes of logic by which deductions are made, but it should be possible for you to "feel" that the writer intends you to "read between the lines."

The ability to arrive at general conclusions and judgments. Again, these may more reliably be arrived at through discussion of the material, but you should be able to draw your own conclusions from what you have read and be able to evaluate or judge the material reasonably accurately. It is something that we tend to do without conscious effort but we shall consider in Chapter 3 how the process may more effectively be conducted consciously and systematically.

The ability to relate knowledge to experience. You should be able to understand allusions made by the writer and the writer should be able, on most occasions, to assume a certain amount of prior knowledge on your part. You should be able to relate what you read to what you have read and learned previously and to modify accordingly, if necessary, your understanding of, or insight into, a subject. In this way your comprehension of what you read will improve over a period of time and the occasions on which you have difficulty in comprehending will be reduced.

Comprehension in reading is rarely constant. There does appear to be an upper limit to the *rate* at which any individual can comprehend, but it is also true that most of us operate at a much lower level than we are capable of achieving. This course is designed to bring you closer to your own upper limit. However, a number of factors can temporarily affect both the speed and quality of your reading comprehension. You may already be aware of some of these factors and it will be useful if we take a look at them so that you can identify the effect each has upon your own reading.

Factors Affecting Comprehension

Speed of perception is an important factor. Some people are quick to grasp, while others are slower. Slower workers often encounter more difficulties than faster workers because of pressures of shortage of time in most jobs and courses of study. It takes them longer to comprehend and they may have to rush in order to keep up. In this way, especially in reading, important points may be missed. Slow comprehension is not necessarily a sign of poor comprehension, but it may well result in this if time runs out before the reading is completed. This course will help slow readers to overcome this problem. Readers who are slow to comprehend also tend to be overconscientious by giving the same amount of time and attention to trivial reading tasks as to important ones. You should aim to give a piece of reading matter just the amount of time and effort it merits, neither more nor less. Becoming more flexible helps you to make the best possible use of your speed in perceiving and comprehending the meaning of the written word.

The accuracy of your perception is another factor that affects reading comprehension. The most obvious example in which inaccuracies can lead to poor comprehension is where negatives are not seen. Imagine the confusion if we were to read the Ten Commandments without being aware of the negatives. Other inaccuracies can produce other errors which, while perhaps not so dramatic, can be just as serious in their long-term effects. Inaccuracy invariably leads sooner or later to inefficiency.

Memory and the ability to recall information once read can affect comprehension. It is possible to understand without being able to remember much of the actual content of the material but it is much more useful if information and ideas from the material can be clearly remembered. Good retention makes it easier to defend one's own interpretation of the material and to locate any points that require checking.

Your **motivation or purpose** in reading is important. You must *want* to read and have a clear knowledge of your purpose in order to read effectively and be in a receptive frame of mind for the information and ideas being expressed. Some readers are easily distracted from their reading, while others can read well even in a busy office where people are talking or using typewriters.

Concentration (see also page 169) can be affected either by internal sources of distraction, such as anxieties or wandering attention, or by external noises and movements. A high level of concentration is necessary for good comprehension and it affects many of the other factors that are being discussed here. **The level of difficulty** (see also pages 98-100) and **the nature of the material** can affect comprehension. You cannot do a great deal to control these factors but they can help or hinder you and so affect your efficiency in reading.

Among the other factors that affect both the speed and quality of reading comprehension are **the ability to anticipate when reading** (see also pages 168-169), **vocabulary** (see also page 167), **your general background of knowledge and experience, and the ability to read critically.**

Speed and quality interact and you will soon discover that if you try to read *too* quickly at any point in this course the quality of your comprehension will suffer. You will also find that a high score for quality of comprehension gives you the confidence to press for further increases in speed. A careful balance between speed and quality is essential for achieving greater efficiency in reading.

Improving Comprehension

Now that we have attempted to define reading comprehension a little more closely and have discussed some of the factors that affect it, we can examine some of the ways by which the quality of reading comprehension can be improved. We shall consider six methods.

1. *Your comprehension will be improved by trying to improve separately each of the elements that affect it.* Speed and accuracy of perception improve as a natural result of gradually trying to read faster and of being tested as you do this. Retention of information, as we have already said, improves as a result of being tested on what you read, especially when, as in this course, part of most tests consists of questions that cannot be answered

unless you have remembered well what you have read. A sense of purpose can be developed by taking the trouble to define your reasons for reading before you begin to read. Concentration can be improved by trying to do as much of your reading as possible, and certainly your most important reading, where you know from experience you can concentrate well. Anticipation skills can be improved by trying to relate what you are reading to what you have just read in order to be able to see the general direction in which the writer is heading. Vocabulary can be improved by getting into the habit of using a dictionary to look up unfamiliar words. Most of these points will be developed more fully in later chapters.

2. *Test the quality of your comprehension regularly and in a variety of ways.* This in itself will produce improvements and, even with the material you read between the times you spend working through this book, you should try to test comprehension by means of:

a) **Summary.** Try to condense the material to its essentials and refer back to the original to make sure you have omitted nothing.

b) **Questions.** Try to ask yourself questions about the material or ask someone else to pose the questions. Check your answers by reference back to the material.

c) **Discussion.** We shall have more to say about discussion in a moment.

3. *You will improve your reading comprehension if you read critically.* Chapter 3 will suggest ways in which this can be done.

4. *The use of study techniques where appropriate helps to improve comprehension.* Chapter 4 will suggest how reading materials may be studied more systematically and effectively.

5. *The value of discussion, not only as a method of testing comprehension but also as a method of improving it, is considerable.* This is probably the best, though the least precise, method of assessing the quality of reading comprehension. If

you discuss what you read with others, you can compare your own understanding and interpretation of the material with theirs and modify your assessment if you realize they have grasped important points which you have missed. If you are a member of a tutorial group that is using this book, opportunities for discussing reading material should be available frequently. If you are working through this book on your own you should use every opportunity that arises for discussing your reading with others who have also read the material. Discussion is the best way of improving comprehension because your understanding and interpretation are immediately either reinforced or challenged. Discuss ideas, generalizations, main points, and abstract ideas because it is through these that the greatest possibilities for developing comprehension skills lie. You will find, however, that discussion also helps in improving the retention of information because facts are better remembered in association with the ideas or principles to which they are related. It is very difficult to remember facts in isolation. So discussion not only helps you to understand the material, it also helps you to remember it better and for longer.

6. *Finally, wide and varied reading is an excellent method of improving reading comprehension.* Perhaps the most effective way of achieving long-term improvements in your reading abilities is, in fact, to read widely and to discuss with others what you have read. If more people did more reading and discussed their reading more fully and more searchingly, there would be little need for courses such as the one offered in this book.

An additional aid to these methods of improving comprehension lies in improving your other language communication skills—thinking, listening, speaking, and writing. There is some evidence to suggest that when you improve your ability in using one communication skill, you help to improve your use of the other skills. Improvement in listening ability, particularly, assists improvement in reading because listening, too, is receptive in nature.

DIFFERENCES BETWEEN READERS

You learned at the beginning of this chapter that one of the basic differences between slow and fast readers is the extent of the eye span, or **span of perception**. Other basic differences are that a slow reader **"regresses"** (goes back to read something again) frequently, and that he tends to **"subvocalize"** (says each word silently to himself as he reads).

There are several other differences that are worth mentioning. Generally speaking, *the slower reader has a narrower vocabulary range than the efficient reader.* He frequently comes across words he does not understand and, because he cannot guess the meaning accurately enough from the context and has not acquired the habit of using a dictionary, his inefficiency is never remedied, and, indeed, gets worse.

Because he regresses frequently, *the slower reader is not able to achieve any degree of rhythm in his eye movements* as he progresses along the lines and down the page. Thus his reading is jerky, uncertain, and ineffective. Reading rhythmically is relaxing and assists both concentration and the ability to anticipate what is coming next in the material. Rhythmic eye movements come naturally to the efficient reader and you will find that by the time you reach the end of this book your own eye movements are much more rhythmic and confident. You should not, however, try deliberately to make your eyes move rhythmically. Allow this to develop naturally as a result of reading faster.

A slow reader is often unsure of, or not sufficiently precise in defining, his purpose in reading. He simply thinks either that he reads something because he has to, or, during his leisure time, because he wants to be entertained. Reading situations are rarely as simple as this. He needs to learn to ask *why* he has to read the material or *why* he expects to be entertained by it. We shall return to this point in Chapter 3.

Perhaps the most important difference between the techniques used by slower readers and those used by faster, more efficient readers *lies in the nature and extent of the individual's knowledge and experience* (particularly experience in reading) and in how thoughtful and critical is his approach toward them. The wider his knowledge, the better he is able to comprehend new material; the more varied his experience in reading, the surer and more effective his handling of it. Efficiency in reading comes largely from practice, from variety in the content of the material read and from variety in the level of difficulty of the material read. The really efficient reader enjoys reading for its own sake and particularly enjoys reading material that challenges his ability to comprehend and to interpret and thus further enlarges his knowledge and experience.

Other important differences will become apparent later, but for the moment let us take a closer look at methods of avoiding regressions in reading.

AVOIDING REGRESSIONS

There is really only one way to avoid having to go back to read something again and that is simply to *refuse to allow yourself to go back.* You will find at first, quite naturally, that your comprehension suffers, but if you persist in refusing to regress you will soon find that the quality of your comprehension rises back to its normal level and will very probably improve. *Follow this rule in all your future reading of the exercises in this book* and when you reach the point at which your regressions are rare rather than frequent (as they probably are at present), you will find that the number of regressions you make in your normal daily reading has also decreased markedly.

Use a sheet of paper, or a postcard, or cover the material you have read with your hand to help you break the habit initially. Draw the paper, card, or hand down over the page as you read. You can also use this device as a pacing aid to help you increase your reading speed gradually. You should be able to avoid regressions without this aid within two weeks. This is how you will cure yourself of the habit of regression, but let us examine some of the reasons why we regress in our reading and also what happens when we try to eliminate regressions as far as we can. This may help to give you the confidence to make the essential break with habitual regression.

Why We Regress

Regression occurs in adults' reading largely through **habit,** and its origins probably lie far back in early school days. The number of occasions on which anyone *needs* to regress is, on most materials of an average level of difficulty, very low. So it is ourselves and our longstanding habits we should blame and not the material we are reading. The habit develops primarily from a **lack of confidence** in our own reading abilities rather than from an inability to understand what a writer is trying to communicate. If we break the habit by deliberately refusing to regress, we soon find how much we have been relying on regressions unnecessarily. Other reasons for regressions are, of course, the **level of difficulty** of the material, an **unwillingness to take more care and trouble over our reading,** and a **lack of concentration.** These reasons will always necessitate occasional regressions.

The Effects of Reducing Regressions

Now what happens when we try to reduce regressions to a minimum? We have already made the point that, **for a short time, quality of comprehension suffers.** Once the initial period of difficulty is over and the brain begins to work as it should, **many skills and processes involved in reading begin to improve. Concentration improves,** because previously, knowing we could always regress, we have in fact only been half concerned with the act of reading. Our attention wandered because we knew we could take the lazy way out.

The most important effect of reducing regressions, however, is **an increase in the speed of reading.** This can often be quite sudden and dramatic, with increases of more than 100 words per minute on material of average difficulty. **The ability to read rhythmically,** letting the eyes sweep along the lines in light, easy movements, is improved because of the removal of the main cause of irregularity. **The ability to anticipate** what is coming next and even later on in the material is improved because the improvement in concentration and the increases in reading speeds make it easier to grasp the overall pattern of construction of the material. It is not surprising that, since many of these improvements occur within a very short space of each other, there is **an increase of confidence** in one's own ability to read rapidly and efficiently.

A NOTE ON SUBVOCALIZATION

Subvocalization (sometimes described as "inner speech" or "reading aloud silently") cannot be reduced in the same easy and dramatic manner, nor is there any evidence that it limits reading speeds unduly. It is simply a nuisance. It has been found, in fact, that it is possible to read aloud at speeds up to 475 words per minute. So it should be possible to read as fast as this, if not faster, when subvocalizing. Subvocalization is similar to **vocalization,** in which a person is unable to read silently at all. The latter is a more fundamental reading disability and really requires treatment in a reading "clinic." The local education authority should be able to advise the whereabouts of the nearest one (some colleges and universities provide courses for very poor adult readers), though there are not many in some states. However, *the best "cure" for subvocalization, if the reader wants one, is simply to forget about it.* As higher reading speeds are reached through practice, subvocalization becomes less and less pronounced, though it may never disappear altogether, even when skimming.

Some other elementary points to watch for in your reading are these:

1. Avoid moving your head from side to side as you read.
2. Do not follow the words with a finger or a pencil in an attempt to speed yourself up. It does not work and will, on the contrary, tend to slow you down.

3. Avoid drumming with your fingers on the table as you
    read. This, and similar irritating habits, do not help.
    They merely indicate the presence of too much nervous
    tension. Relax.

EXERCISES

*Try to read this next set of exercises faster than you read the last
set.* Try to see the passages in terms of **groups of words** rather than
single words. Be more concerned with the *meaning* of what the
writers are trying to communicate to you than with studying
too closely the actual words they use. Remember that you are
following a method already successfully followed by hundreds
of students and that as they succeeded so will you—if you have
confidence in your own abilities and the necessary amount of
determination to follow the instructions given.

Instructions

Read through the passage **once only** *as quickly as you can
without loss of comprehension.* **Do not regress.**

Try to improve upon your previous best performance.

Begin timing and begin reading NOW.

IT BEGAN WITH THE WOMEN—I

by Edward Crankshaw

Those who took part in the great revolution which was
to sweep away the Russian Tsar and change the course of
history had no idea that they were caught up in an earth-
quake; they thought they were demanding better government
and bread. Those who watched from the outside were pleased
that they had taken action and wished them luck. It was a
most misleading overture to chaos, horror, the installation of

renewed tyranny of unimagined rigour. Nearly 40 years were to pass before anyone could be certain that the children of those revolutionaries, and their children, were at last escaping from the consequences of the process so innocently set in motion in March 1917.

It began for all practical purposes on 8 March 1917, and it was all over, people thought, with the abdication of Nicholas II eight days later. But this was only a beginning. In fact it was not over until October. And 25 October, the day of Lenin's revolution, though it marked the victory of Bolshevism, was only the overture to a far more terrible time: two years of atrocious civil war, complicated by intervention from the outer world, and followed by famine.

The year 1917 opened with defeatism. The Empress's confidant, Rasputin, was dead, but it made no difference. The Empress went into holy communion with his soul and Nicholas locked himself up in virtual retreat with his family among the splendours of Tsarskoe Selo just outside Petrograd (now Leningrad). In the city people froze and starved and murmured while conspirators devised lunatic schemes. The vast armies, reduced by desertions, lay out in their trenches in the forests and the frozen wastes, still holding down 160 divisions of the Central Powers. Everything was in suspense.

On 8 March, the Tsar pulled himself together and set out once more for his headquarters at Mogilev, behind the Front. On that day, too, the revolution started. It began, not with banners, not with an organized march of workers, not with a mutinying garrison, certainly not with a call to arms by professional revolutionaries; it began with the women, who started demonstrating outside the empty shops of Petrograd. Protest meetings formed like magic. Soon there were processions marching down the great avenues with the immemorial Russian wail: "Give us bread!"

Workers downed tools and joined them. A few companies of soldiers, very few on that first day, broke barracks and joined in too. The demonstrators were surprised, almost shocked to find themselves among so many. When darkness fell they drifted off to their dark, cold homes, wondering about tomorrow.

430 words

Write down the time taken to read this passage and then attempt the Comprehension Test.

COMPREHENSION TEST

*Do **not** refer back to the passage.*

A. Retention

    1. Name the Empress's confidant.

    2. Where were the Tsar's war headquarters?

    3. When did the revolution really begin for all practical purposes?

    4. What was the cry of the protest marchers?

    5. Who joined the demonstrators?

B. Interpretation

    6. How aware were "those who took part in the great revolution" of the consequences of their actions?

    7. Why was the revolution "a most misleading overture"?

    8. What impression is given of the first day of the revolution?

    9. What was the nature of the relationship between the Empress and her husband, Nicholas II?

    10. Why did the demonstrators wonder what would happen next?

Convert the time taken into "words per minute" by means of the Reading Speed Conversion Table on page 254. Enter the result on the Progress Graph on page 256.

Check your answers to the Comprehension Test against the answers given on page 257. Enter the result on the Progress Graph on page 256.

C. Discussion—Discuss one of these questions (orally if in a group, in writing if studying alone). *You may refer back to the passage.*

11. Do most revolutions begin in the way the Russian Revolution began?

12. How influential was Rasputin? Why?

13. Why was Petrograd renamed Leningrad after the revolution?

14. Do you think that the people deliberately waited until the Tsar had left Petrograd before they demonstrated?

15. What indications are there in the passage that the Russian winter is long and cold?

Note

There will be further passages on this subject as you proceed through the book so that you will be able to compare your performance from time to time with your performance on this exercise.

Proceed to the next reading exercise.

Instructions

Read through the passage **once only** *as quickly as you can, without loss of comprehension.* **Do not regress.**

Try to improve upon your previous best performance.

Begin timing and begin reading NOW.

CAN MAN SURVIVE?

by Joyce Eggington

You are walking down a long dark tunnel which gets narrower and hotter and louder. The clamour of many voices and machines makes it impossible for you to think; the air seems too oppressive to breathe. You feel yourself enclosed in too little space from which there is no escape; you can go only in the direction of the crowd, which becomes more and more claustrophobic. It is a nightmare, except that you are awake.

For this unpleasant experience, thousands of New Yorkers are paying a dollar a head and are expected to continue doing so in increasing numbers, for the next two years. The event is a unique exhibition at the American Museum of Natural History, the first of its kind in the world. It is called *Can Man Survive?* and it is a serious attempt to pose that question.

In the early stages the exhibition is a delight. One enters a broad, carpeted hallway from which all exterior sounds are excluded. It is cool and softly lighted; there is an impression of much space, and the sound of birds singing. The world of nature is projected on film, and a commentator gently drives home the message that for millions of years life has been maintained by a self-regulating natural balance between the amount of food and oxygen available, and the number of animals which survived.

The point is made that nature's methods are sometimes harsh: one beast devours another. From then on the corridor narrows and darkens as films and photographs trace man's advance into the industrial age. At first the positive side of this is shown, but then the tunnel gets even narrower, the background noise of machines, factory furnaces and automobiles becomes noisier. There are film flashes projected from every wall, crowding in on the visitor, with the impression of conveyor belts, tractors and teeming masses of people.

Amid the noise there is a shot of a new-born baby, and the visitor is immediately attracted to it as the one thing endearing and innocent. But a voice from somewhere in the tunnel comments: "This baby's chances of survival are excellent. So are every baby's. There are twice as many mouths to feed as there were 40 years ago, and already there is not enough food to go round. . . ."

By the end of the exhibition the tunnel has become so narrow and oppressive that the visitor almost stumbles over a beat-up bus and a pile of garbage cans which half-blocks the exit. There, written in red on the wall, is the message "Can Man Survive? It is up to you."

After such bombardment of the senses it is a sweet relief to walk out of the museum into the Manhattan traffic.

The scientist who planned the exhibition is Dr. Harry L. Shapiro, chairman of the museum's Department of Anthropology. He feels that the picture as presented is not in the least overdrawn. "I think the prospects for survival are very grim," he said in an interview. "Our only hope is to make people aware of it in time, because in the last analysis nothing can be done to regulate the consequences of our technology unless people are conditioned to accept regulation. A government cannot tell a family that it must have no more than two babies and one car. Instead we have to make people understand that survival depends upon their cooperation."

Of all the threats to survival—air and water pollution, food shortage, pesticides and noise—Dr. Shapiro believes that by far the greatest is the population explosion. "This problem is overwhelming, and we do not have much time to solve it," he said. "Already we have a situation far more insidious and far more dangerous than the atomic bomb. At least for the moment that is under control. But the population explosion is getting bigger day by day, and the prospects of getting an international consortium on this are not good."

Dr. Shapiro said that if he had an entirely free hand there was one further point he would have had the exhibition make. "There used to be so many hazards to man's survival that it was necessary for him to produce numbers of children. Now we—the developed nations—have created a terrible problem for the under-developed people. We have gone into their countries and killed off the bugs that were decimating them. We have changed the entire pattern of living for them, while they have not changed theirs one scrap. But for all sorts of reasons we have still not succeeded in showing them how to adjust to the new pattern by limiting their reproduction rate. So in a generation or two we may have a major disaster—a famine or a revolution which could irrevocably affect the entire world."

800 words

Write down the time taken to read this passage and then attempt the Comprehension Test.

COMPREHENSION TEST

Select the most suitable answer in each case.

Do **not** *refer back to the passage.*

A. Retention

1. The "Can Man Survive?" exhibition took place in:
   a) London.
   b) Paris.
   c) Washington.
   d) New York.

2. There are now twice as many mouths to feed in the world as there were:
   a) 4 years ago.
   b) 40 years ago.
   c) 140 years ago.
   d) 400 years ago.

3. The message at the exit from the exhibition read:
   a) "There are too many children in the world."
   b) "Please mind the garbage cans."
   c) "Can Man Survive? It is up to you."
   d) "Can man survive? Not without birth control."

4. Dr. Shapiro believes that by far the greatest threat to human survival is.
   a) the population explosion.
   b) air and water pollution.
   c) food shortage.
   d) pesticides.

5. Dr. Shapiro said that, as a result of our failure to show underdeveloped people how to limit their reproduction rate, in a generation or two there may be:
   a) more people in the world than land for them to live on.
   b) a plague or an epidemic to restore the balance of nature.

    c) a considerable increase in the birth rate.

    d) a famine or a revolution that could irrevocably affect the entire world.

B. Interpretation

6. "... you can only go in the direction of the crowd, which becomes more and more claustrophobic." Claustrophobia is:

    a) a dread of confined spaces.

    b) a state of intoxication.

    c) a fear of heights.

    d) a fear of the dark.

7. The world of nature, as portrayed in the early stages of the exhibition, is:

    a) idyllic.

    b) cool.

    c) noisy.

    d) lifeless.

8. At the end of the exhibition the "beat-up bus and a pile of garbage cans" symbolize:

    a) the world of nature as it is today.

    b) the failure of man to develop beyond the primitive stage.

    c) the condition of Western industrial civilization.

    d) the problems facing the underdeveloped countries.

9. Dr. Shapiro proposes that, to secure people's cooperation in limiting the size of their families, governments should:

    a) tell them they must have no more than two babies.

    b) use persuasion.

    c) form committees to look into the problem.

    d) use force.

10. The most important single point made by the exhibition was that:

    a) the answers to man's problems were simple ones.

    b) there was no hope for man's survival.

    c) it was up to governments to get together if man was to survive.

    d) it was up to the individual to act if man was to survive.

Convert the time taken into "words per minute" by means of the Reading Speed Conversion Table on page 254. *Enter the result on the Progress Graph on page* 256.

Check your answers to the Comprehension Test against the answers given on page 258. *Enter the result on the Progress Graph on page* 256.

C. **Discussion**—discuss one of these questions (orally if in a group, in writing if studying alone). *You may refer back to the passage.*

11. Do you feel that the picture of man's predicament, as presented by this exhibition, is an accurate one?

12. Should governments dictate what must be done to solve the problem of man's survival or should they try to persuade people to give their support to the necessary policies?

13. Why do you think Dr. Shapiro did not have "an entirely free hand" in mounting this exhibition?

14. *Can* man survive? What do you think?

15. Will it take "a famine or a revolution" to make man act to save himself?

Proceed to the next reading exercise.

Instructions

Read through the passage **once only** *as quickly as you can without loss of comprehension.* **Do not regress.**

Try to improve upon your previous best performance.

Begin timing and begin reading NOW.

THE MAN WHO BUILT LIBERIA
by Bobby Naidoo

William Vacanarat Shadrach Tubman, late President of Liberia in West Africa, was the most outstanding man in the chequered history of Africa's first republic. In spite of abortive assassination attempts, *coups d'état*, and some rather unsavoury local politically-tainted trials of would-be usurpers of power, Tubman weathered everything for 25 years as President with a benign smile and a puff of one of his Churchillian cigars, which were regularly shipped to him from London.

He had a ready answer to all the attacks made on Liberia: its flag of convenience, which now has the largest registered maritime tonnage of some of the most modern shipping in the world; doubts about its diamond fields and the smuggling of diamonds across the border from Sierra Leone, which earns the Government a valuable revenue of several million dollars a year; corruption in the police and civil service where the well-known West African "dash" (tip) can buy favours; and the charge that the country is nothing more than a colony of the United States of America.

But physical development of Liberia by its own efforts under Tubman is there for all to see—achieved, as Tubman liked to point out, without the assistance of British colonialisation, which has benefited so many other now independent countries in Africa. He masterminded the whole of his national scene from a $16 million Executive Mansion in Monrovia, the capital, complete with nuclear-proof shelter, private chapel, luxuriously appointed presidential apartments and entertaining suites where 1,000 guests can be seated at a full state banquet with the finest foods imported from all parts of the world—salmon flown from Scotland, caviar from Russia, steaks from Texas, tomatoes from the Canary Islands, and wines from France, Germany and Italy—nothing but the finest will ever do.

The Mansion stands on the shoreline overlooking the capital in expansive gardens against a giant-sized pool. It has its own water supply and power plant and all is watched over by a colourfully uniformed Mansion Guard under a commanding officer, whose own uniform bristles with gold braid and a wide array of medals.

Liberia's governmental system is based on the American pattern, and calls for a great deal of local "politicking." All leading officials are political appointees responsible for their areas' loyalties to the Head of State. Tubman left nothing to chance. He checked up personally, travelling thousands of miles every year, visiting remote regions of his 43,000 square mile country. Political favours and jobs for the boys can extend from clerical posts to appointments in Liberia's vast diplomatic network and other well-paid posts with international organisations and agencies of the United Nations.

They might cover things such as full first-class travel and medical attention and treatment in leading private clinics in Europe or the USA for a loyal but ailing party supporter, and full education scholarships for his children. A thousand Liberian students were regularly studying abroad at any one time during Tubman's reign of office—about 150 of them in Britain—nearly all paid for under his general national scholarship programme. Some wealthy Liberians whose incomes are derived from well-paid government positions, rubber plantations they own, or local directorships of international companies operating in Liberia, pay for their sons and daughters to go to England's most expensive schools.

An African nationalist, who has vigorously pursued an open-door policy directed towards the economic development of his country, President Tubman was an activist. He visited the United States for economic discussions. He visited Liberia's prosperous neighbours in West Africa, the Ivory Coast and its President Houphouet-Boigny, and flew across the continent to visit Kenya for discussions with President Jomo Kenyatta. In April 1968, he was chairman with Leopold Senghor of Senegal as his vice-chairman, at a conference of 11 African countries which agreed to create a West African Common Market.

National development in the last 25 years is staggering compared with what happened after 1822, when freed Negro slaves first landed on the West African coast as a result of American philanthropy. When Tubman took office the national revenue of the country was barely a million dollars. Today it's more than $60 million. Virtually single-handed, but later assisted by a growing band of foreign-trained Liberian economists and specialists, he drove Liberia into the twentieth century as an African nation which benefits from large-scale agricultural, mineral and commercial projects operated by American, Spanish, German, Italian, Dutch, French, Israeli and British companies.

Liberia's international reputation has been based on rubber, but iron ore has been a leading factor in the national economy since 1951, when the Liberian Mining Company first went into production in the Bomi Hills, north of Monrovia. Deeply involved in the financial structure and management of vast mines at Mount Nimba are the Grangesberg Company of Sweden and Bethlehem Steel of America, but the Liberian Government owns 50 per cent of the operation. August Thyssen, Dortmund Hoerder, Friedrich Krupp and Rheinische Stahlwerke are involved in the Bong Mining Company, the biggest overseas mining operation ever undertaken by the German steel industry.

The mining companies have provided widespread employment, new building developments, roads and railways linking inaccessible parts of the country. US Aid funds have provided Liberia since 1945 with a hydroelectric project, a national telecommunications system, a highways system, national electrifications and urban development.

President Tubman was born at Harper, in the southern-most country of Maryland in Liberia. His father, Alexander Tubman, a Methodist minister, was a descendant of the early settlers from the USA, but his mother, Elizabeth Rebecca Barnes, came from the southern American State of Georgia. The young "Shad"—his intimates always called him that—went to school at the Methodist Seminary in Harper. He then studied law, was admitted to the Liberian Bar at 22, was made County Attorney at 25, served as a colonel in the Liberian Army and in 1937 was appointed Associate Justice of the Supreme Court by the then incumbent President Edwin Barclay. Tubman married President Barclay's niece, who was educated in Europe, and became 18th President of Liberia in 1944 at the age of 49. He was an Anglophile, had his suits tailored in Britain, sent his sons to a leading public school in Britain, and has always encouraged his scholarship grantees to study in England.

But British business, which virtually controlled the Liberian economy before World War II, is now in third place having only 8.4 per cent of Liberian exports compared with the USA's 45.6 per cent and 9.3 per cent of Liberian imports compared with USA's 44.6 per cent and Germany's 12.6 per cent. There has been a tremendous development of US trading through Lebanese businessmen who have a strong hold in commerce with sales franchises in automobiles, heavy equipment and consumer goods. Liberia has close links with Israel and there is also a growing band of Indian businessmen dealing mainly with goods from the Far East, Hong Kong and Japan.

Britain was the first country to recognise Liberian sovereignty in 1847 during Queen Victoria's reign and gave the new nation a gunboat. Now America is the leading influence in an African republic that strives to make its independent voice felt in African and world affairs.

1200 words

Write down the time taken to read this passage and then attempt the Comprehension Test.

COMPREHENSION TEST

Select the most suitable answer in each case.
Do **not** *refer back to the passage.*

A. Retention

1. President Tubman had been in office for:
    a) 2 years.
    b) 5 years.
    c) 25 years.
    d) 52 years.

2. President Tubman's Executive Mansion cost:
    a) $6 million.
    b) $16 million.
    c) $26 million.
    d) $36 million.

3. Freed Negro slaves first landed on the West African coast in:
    a) 1822.
    b) 1832.
    c) 1842.
    d) 1852.

4. President Tubman's nickname was:
    a) "Tubby."
    b) "Bill."
    c) "Shad."
    d) "Pres."

5. In the Liberian economy, British business is in:
    a) third place.
    b) first place.
    c) thirteenth place.
    d) tenth place.

B. Interpretation

6. A *coup d'état* is:
    a) a battle.
    b) an attempt to overthrow a government.
    c) a revolution.
    d) an unsavory politically tainted trial.

7. An open-door policy means that a country will:
    a) trade with any other country.
    b) only accept American investment.
    c) not sign treaties with anyone.
    d) only trade with African countries.

8. In the last 25 years the Liberian economy has:
    a) hardly developed at all.
    b) developed slowly.
    c) developed steadily.
    d) developed rapidly.

9. The closeness of the relationship between Liberia and the USA may have been helped by the fact that:
    a) President Tubman was born in America.
    b) President Tubman's ancestors were among the early settlers from the USA.

   c) The USA dominates the economies of all African
      countries.
   d) Britain's only contribution to Liberia's development
      was a Victorian gunboat.

10. British influence in Liberia is probably:
   a) declining rapidly.
   b) greater than the trading figures indicate.
   c) greater than the influence of the USA.
   d) nonexistent.

*Convert the time taken into "words per minute" by means
of the Reading Speed Conversion Table on page* 254. *Enter the
result on the Progress Graph on page* 256.

*Check your answers to the Comprehension Test against the
answers given on page* 258. *Enter the result on the Progress Graph
on page* 256.

C. **Discussion**—discuss one of these questions (orally if in a group,
      in writing if studying alone). *You may refer back
      to the passage.*

11. Do you think the Liberian government could best be
    described as stable or unstable?

12. What general impression do you form, from your reading
    of the passage, of President Tubman and his government?

13. What is African nationalism? What do its adherents believe?

14. Why do you think President Tubman liked to describe
    Liberia's flag of convenience as "a flag of necessity"?

15. What factors do you feel influenced the decline of British
    business with Liberia and what prospects do you feel
    there are for a revival?

PRACTICE FOR THE COMING WEEK

Spend about **half an hour each day** reading your daily newspaper.
Each day, try to increase the amount you can read in this time.
To test your comprehension, select one article each day. Try

either to write down as many of the facts given as you can remember (if it is a news article) or to write down the main points the writer is making (if it is a feature article). For a good level of comprehension, you need to remember about 70 percent of the facts, or most of the main points.

Note

Cut out the articles you use for comprehension testing and keep them with your answers, in a file. Keep a record of your results and compare them from time to time with your results at the beginning of the course to see what progress you are making.

You will recall that before you took the Initial Tests, you spent a few seconds looking through, or *previewing*, them. Get into the habit of similarly previewing everything you have to read. **Count this time as part of your reading time.** This applies both to your normal, daily reading and to your practice reading.

You should also acquire the habit of *reviewing* what you have read, by means of another rapid skimming, to check that nothing of importance has been overlooked and to enable you to fix important points more securely in your mind. **This *preview-read-review* approach is a technique often used by naturally efficient readers.**

Avoid regressions when reading.

CHAPTER SUMMARY

1. **Nature of the Reading Process**
   a) The eye can only read if allowed to fixate momentarily on words and phrases between *saccades* or movements.
   b) How quickly a person can read depends on, among other things, the width of his eye span (how much he sees at each fixation).

2. **Nature of Comprehension**
   a) Comprehension includes the ability to retain information and recall it when required.

b) Comprehension includes the ability to select important points.
c) Comprehension includes the ability to interpret information and ideas.
d) Comprehension includes the ability to make deductions from what has been read.
e) Comprehension includes the ability to arrive at general conclusions and judgments.
f) Comprehension includes the ability to relate knowledge to experience.

3. Factors that Affect Comprehension
Some of the main factors are:
a) speed of perception.
b) accuracy of perception.
c) memory and the ability to recall information.
d) motivation or purpose.
e) concentration.
f) level of difficulty of the material.
g) nature of the material.
h) ability to anticipate.
i) vocabulary.
j) general background of knowledge and experience.
k) ability to read critically.

4. How Comprehension Can Be Improved
a) Try to improve each of the factors that affect it.
b) Test the quality of comprehension regularly and in a variety of ways.
c) Read critically (see Chapter 3).
d) Use study techniques where appropriate.
e) Discuss reading materials as often as possible.

5. Differences Between Readers

Inefficient Reader	*Efficient Reader*
a) Narrow eye span	Wider eye span
b) Regresses habitually	Reduces regressions
c) Subvocalizes	Tends not to subvocalize
d) Restricted vocabulary	Wide vocabulary

	Inefficient Reader	*Efficient Reader*
e)	Irregular eye movements	Rhythmic eye movements
f)	Lacks purpose in reading	Purposeful in reading
g)	Limited background of general knowledge and experience	Broad background of general knowledge and experience

6. Avoiding Regressions

a) In practice sessions, refuse to regress for any reason at all.

b) Use a sheet of paper, a postcard, or your hand to place a physical barrier against regressions.

THREE

Purposes and Materials

DEFINING PURPOSES IN READING

When we are reading, we are usually at least vaguely aware of our **purposes.** We know that we are reading either because we **want to** or because we **have to** and we are aware of the differences in our approaches for each of these purposes. But we need to be much more specific and exact in our definition and awareness of purpose if we are to deal more effectively with our reading. This is particularly true for reading that we *have* to do (such as reading at work), which is the kind of reading we are primarily concerned with in this book.

In addition to being able to recognize two basic purposes in reading, we know, when we pick up something to read, that we have some **expectation** of what we hope to learn from the material. Even before we read a line, we have a vague idea about what we are going to be told. These basic purposes and expectations always exist but we need to make ourselves much more fully aware of them if we are to be able to adjust our reading techniques to suit them.

One way, in practice sessions, *to develop this more exact awareness of purpose is to make a brief written statement of the reasons*, as far as you can analyze them, *for which you are reading the material and also of your expectations from the material.* For example, let us say you are reading a newspaper. Why are you reading it? For practicing faster and more efficient reading techniques, possibly. You have selected a particular article for this purpose. Why have you selected *this* article? Because (let us say) it is relevant to your work in some way and it seems, from the heading, to offer some new information. So, then, what are your expectations from the material? Let us say that they are that you will learn something new about a subject relevant to your work. Or you may wish to select the main points the writer makes and to note any other important facts and ideas, and in doing this to be able to do it a little faster and a little better than you did the previous day.

This conscious analysis of your motives and purposes in reading will encourage, even over a short period of time, more productive, more confident, and faster reading. If you do it regularly every day, it will soon become automatic, so that after a week or so you will find yourself doing this advance analysis of your approach to material without having to go to the trouble of writing it down.

A Questionnaire for Assessing Purposes in Reading

During the coming week, select two or three pieces of reading matter and use the following short questionnaire to help you to make a written analysis of your purposes in reading:

A. *The Nature of the Material*
   1. What kind of reading matter is this (e.g., newspaper article, chapter of a book, a report)?
   2. How long is the material (the approximate number of words)?
   3. What is the relevance of the subject matter to your work, hobby, or interests?

B. *Purposes in Reading*
   1. Why did you select *this* piece of material?

2. How much information do you want from the material (e.g., main points only, full retention of all facts and ideas, only new information)?
3. Is the information to be put to any specific use (e.g., to help with a problem at work, to add background to a subject you are studying)?

C. *Expectations in Reading*
   1. What, in brief outline, do you expect the writer to tell you?
   2. How useful do you expect the information to be?
   3. Do you expect to find the material easy or difficult to read, interesting or dull, entertaining or serious?

D. *Questions to Ask* after *Reading the Material*
   1. How far were your purposes fulfilled and your expectations justified in reading the material?
   2. Did defining your purposes and expectations beforehand increase your understanding of and insight into what you read?
   3. Were you able to read the material a little faster because you thought about it first?

Careful assessment of purposes and expectations in reading invariably produces improvements if it is done well, and it provides an important part of the necessary basis for reading efficiently.

HOW WRITTEN MATERIALS ARE CONSTRUCTED

An understanding of the general principles governing the construction of written material is also useful to the would-be efficient reader for several reasons. Knowing how a writer imposes a **pattern of organization** on the information or ideas he wishes to communicate enables a reader to define his purposes in reading much more closely, which enables him to select with greater accuracy the technique he will use in reading the material. It helps a reader to use his anticipation skills to better effect and it helps him, for example, to relate the content of each chapter to the general framework of a book as a whole. It thus enables him to select more accurately the important points the writer is making.

In considering the construction of written material, we shall examine it from three standpoints. **First, we shall look at the types of writing** that a reader will encounter. **Then** we shall look at **the way in which written material is structured. Finally, we shall** look at **the usual pattern of organization of some specific kinds of material.**

Types of Writing

There are, basically, four types of writing—description, exposition, argument, and narrative. Most things that we read can be classified under one or other of these headings. Of course, many pieces of writing will be made up of a mixture of two or more types of writing, but one will usually dominate the others. Let us see how these different types require different approaches and how material where these types are mixed should be dealt with.

Description. In this type of writing, the writer is trying to convey to us a visual image or picture by means of words. He is trying to show us what something looks like. If we are to pick out the main points in the description, to "see" the main elements of the "picture," we must look particularly for the words and phrases that will help us to form this picture or image in our mind's eye.

Exposition. Here, the writer is trying to explain something to us or to set out before us the facts about a situation. We must look for, and be able to identify, the steps in the explanation and understand the logic of the order in which the writer gives us the items of information.

Argument. In this, the writer is presenting the case for or against a point of view or a proposed program of action, or is considering the alternatives in any particular problem. We must look for logical development here, for a thesis supported by evidence and reasons. We must be able to select the "pros" and "cons" from the rest of the material.

Narrative. The writer is telling a story or giving us the sequence of events in a situation. We must look for chronological development or for movement from place to place in order to be able to distinguish the stages in the narrative.

Mixed Writing. Here, the writing contains a mixture of the above four elements and we must be able to decide which element is the **dominant** one and adjust our approach to suit it.

It is rare to find a piece of writing in which the mixture is evenly balanced, so that on most occasions our best solution to the problem is to decide which element is the most important in the light of our purposes in reading.

The Structure of Written Materials

Whatever the type of writing, we all know that it is structured according to the same basic principles. **Words** are linked together to form **sentences** and sentences are structured into **paragraphs**. With a journal or magazine article, this is often as far as the construction goes (though **subheadings** may be inserted to improve the layout they are all too frequently little help to the reader), but in a report the paragraphs link together to form **sections** and in a book they form into **chapters**. Most of us are accustomed to regarding the sentence as the unit of thought. However, one only has to take a sentence out of its context in a paragraph to realize that sentences do not necessarily "make complete sense" (as most of us were taught at school). *It is much more useful for the reader to regard the* paragraph *as the unit of thought, with its key or "topic" sentence and its dependent sentences,* which combine to make up the point or the idea that is being communicated. Especially when reading through something for a general understanding of the contents, we should move from paragraph to paragraph, picking out the key sentence in each one and quickly relating other information or ideas around this. As we progress, we should relate the points together to form a general understanding of the whole article, section, or chapter.

Patterns of Organization

We shall have more to say in Chapter 5 about solving some of the problems encountered in reading particular kinds of reading material but it will be useful here to examine the ways in which some kinds of material are organized by writers. For example, most books, in addition to having the material broken up into chapters, have a table of contents. An increasing number of books provide the reader with **chapter summaries,** in which the main points from each chapter are collected together for convenience and as an additional reference point. Many books contain an **index,** which serves for most people as the main reference point in the search for specific information. However, many of us forget that the purpose of an **introduction** is often to give us some guide to the way in which the book is organized and may even indicate how different purposes in reading can be satisfied most effectively. Frequently, a writer will direct the reader's attention in advance to certain chapters or will tell him which he can omit if he requires only a general understanding of the subject. Yet it is surprising how many readers ignore introductions in books and regard them as a sort of "warming-up" section for the writer or regard them merely as an extended acknowledgments section.

In **journal articles,** the thesis is usually stated in the opening paragraphs and then developed in the rest of the article. Here writers usually limit themselves in the number of points they make. Outside the academic world, journal articles are, more often than not, read once only and then filed for future reference, so the writer has to make sure that his readers do not have to look too hard for the information he wishes to communicate. Frequently, the main points are restated in the concluding paragraphs in order to emphasize them. Many journals now preface important articles with a **summary** of the main points the writer is making so that readers with little time to spare can the more easily select the articles they need to read.

In **reports,** if they are well written, the main points being made are brought together in a **summary** section, which should precede the report. If the report covers more than, say, half a dozen sheets, it may also have a **table of contents** and an **introduction** section to help the reader digest the contents efficiently.

Every piece of writing, besides being predominantly one of four types and being structured in parts to produce the whole (sentences, paragraphs, sections, or chapters going to make up the report or book, for example), has a **pattern.** That is to say every writer, deliberately or subconsciously, **organizes** the content of what he wants to say before he sets it down. The more clearly he indicates the basis for his collection and selection of information and how he has arranged it within the pattern of organization he has chosen, the more efficiently his readers can assimilate what he wishes to communicate. In fact, we *have* to be able to see the pattern if we are to become more efficient in our reading. During the coming week, you should look for types of writing and patterns of structure in your practice reading, in preparation for the work in the next chapter of this book.

CRITICAL READING

Everybody likes to criticize. Most people criticize destructively, partly because the ability and necessity to criticize constructively is commonly supposed to be the privilege and concern of the few and partly because they misunderstand the meaning of the word "criticize." *To criticize means not only to find faults but also to find points of merit,* for example in those parts of the material where the writer has expressed his ideas or facts better perhaps than other writers on the same subject.

Many people seem to believe that to consider, analyze, and evaluate what is seen, listened to, and read requires an effort and persistence beyond that of which the average person is capable. This is not so. Everyone has the ability and the right to examine and assess whatever is being communicated to him. In fact, everybody does pass some kind of judgment, albeit unconsciously, on everything he hears, sees, and reads. What we are saying here is that *this unconscious process can be made much more efficient if it is done consciously and deliberately.* We can go further: everyone *needs* to criticize consciously and according to some logical method, particularly the student and the industrial manager. It should be one of our purposes in reading at all times and on all kinds of materials to evaluate critically the information and the ideas that are being presented to us.

Unless we are prepared to criticize what we read, to examine it carefully, and to form judgments on the contents, our comprehension will be handicapped. Anyone can read a newspaper, a letter, a report, or a book and accept its statements or reject them because he agrees or disagrees with them. But few people will accept the opinions or decisions of someone who is not able to explain his reasons for them because he has not taken enough trouble to analyze and evaluate what he has read.

Students, in particular, need to be critical of what they read. Even with regard to textbooks, it is not enough to accept without question the recommendations of teachers and lecturers. What a teacher recommends as generally the best book on a subject may not be the best book for each individual student. This is to say, there may be another book, covering the same ground, which explains points in such a way that *you* can understand the material better. But you are not likely to discover this fact unless you are prepared to be critical. It is highly improbable that there is only one "best buy" when it comes to books on any subject.

Points to Consider

A critical approach enables you to understand more deeply whatever it is you are reading. It gives you valuable insight into what the writer actually says, his intentions in saying it, his methods of expressing himself, and how well he is able to communicate whatever it is he has to say (that is, whether what he is saying is the right way of saying it for you). A critical approach requires some effort, but with a knowledge of the general principles that should be followed this extra work is minimized. These general principles can be applied, with minor modifications, not only to all types of reading material but also to other communication media such as films, television programs, radio, talks, and lectures. They can be summarized as follows:

1. Consider the informational content (facts, ideas, theme, story—whichever applies).
2. Consider the writer's intentions.
3. Consider how the writer treats his subject (his selection, arrangement, and expression of information; his degree of bias or impartiality; the effect of construction upon the effectiveness of his material).
4. Consider the degree of the writer's success (in the light of his intentions in writing and your own purposes in reading).

Questions to Ask

To put these principles another way, when you have read the material, ask yourself the following questions:

1. a) What exactly does the writer say?
   b) What are the most important points he is making?
   c) Are his facts correct, as far as you can tell?
   d) What are his qualifications for saying what he does?
2. a) What is the writer trying to achieve?
   b) Are his intentions "honorable" (or is he, for example, using tricks of argument or a careful selection of information to influence you unduly)?
   c) Is the material worth reading in detail or is it sufficient to your purpose if you skim through it?
   d) For whom is it being written?
3. a) How does the writer treat his material?
      i) On what basis has he selected the information he presents to you?
      ii) How has he arranged his material?
      iii) Within his arrangement, how has he ordered the points he makes?
      iv) Has he omitted any important items of information?
      v) How effective is his use of language in expression and has it influenced you unduly?
   b) Is the treatment given suitable to the material, and what effect does it have on your acceptance or rejection of what the writer says?
4. a) What is the degree of the writer's success in informing (or convincing) you, bearing in mind:
      i) the readership for which he intended the material?
      ii) his intentions in writing it?
      iii) your purpose in reading it?
   b) If he fails, where, how, and why does he fail and how could the material be improved?

The Informational Content

The questions "What exactly does the writer say?" and "What are the main points he makes?" may seem at first too obvious and unnecessary to ask. However, when you consider, from your own experience, examples of letters, reports, and newspaper accounts you have read and discussed with others, there have surely been times when you have disagreed about the interpretation of certain facts or ideas or even about actual figures quoted and had to reconsider the original carefully to make sure. Even then you may still have had to argue in defense of your own interpretation with the colleagues or friends with whom you were discussing the matter. The contents of a single letter or statement in the press are often open to numerous interpretations, so the answers to these questions are not always obvious and they are therefore worth asking, if only to check that nothing important has been overlooked. A good and quick way of finding the answers is to make a brief list of what you think are the main points the writer makes or the main items of information that he gives and then check this list against the passage.

The answer to "Are his facts correct?" may not be easy to find. Most of the time we have to take a writer on trust and accept what he tells us as being factual until we can check the information with another, perhaps more reliable, source. In a textbook, for example, we often have to accept what we are told because, as students, we are not usually in a position to challenge what a much more experienced person is telling us. It is only when we become more experienced ourselves that we can assess accurately the validity of a writer's information. This problem also faces many managers, who rely on expert information from specialists in making their decisions. This is not to say that the question should not be asked by students and managers. If we do not get into the habit of doubting to some extent what our textbook writers and report writers tell us, we shall find it very difficult to begin being critical later on when we have become too accustomed to absorbing large amounts of material unchallenged. So you should try to check the validity of the information given by, wherever possible, reference to other sources. This will not be possible in every case because of the amount of time that it takes. But where a writer is dealing with a topic on which it is important for you to be sure the information is correct, some cross-checking is necessary.

In considering the informational content of what you read, you should also ask yourself, "What are the writer's qualifications for saying what he does?" These may be indicated, in the case of a book, on the title page or somewhere in the introduction. Before you can really assess the validity of what someone is telling you, you need to know whether or not he is qualified to write about the subject. Your evaluation of his information will be considerably affected by this. You should remember, however, that a writer's qualifications include more than any university degrees or diplomas he may or may not hold in the subject and a writer who has no "paper" qualifications may still be an authority whose information and opinions should be carefully noted. What he actually says and the way he says it, measured against your own experience of other writers' works, is perhaps a better guide than any degrees or diplomas he may have obtained several years prior to writing his book. Get into the habit of discussing books with others and you will be surprised how quickly your ability to pick out the reliable sources of information improves.

The Writer's Intentions

Asking "What is the writer trying to achieve?" may not be such an important query where the material is objectively factual, but it is essential to ask it of advertising or propagandist (in the widest sense) writing. Failure to do this may result in the acceptance of a statement as fact when it is really an inference or a value judgment, both of which are open to opinion and discussion.

You should also ask, in support of this question, "Are his intentions honorable?" If the writer's aim is to conceal certain facts by revealing others, for example, the answer here is necessary for you to form a reasonably accurate opinion about his material. The answers to both questions will be of great help in assessing most writing and may in some cases (for instance, political propaganda) be the main basis of a rejection of its proposals or conclusions. This may lead you to ask, "Is the material worth reading in detail or is it sufficient to my purpose if I skim through it?" Obviously, you do not want to spend valuable time reading biased and prejudiced material when you may know of another writer whose approach seems to be more objective. However, if your purpose in reading in the first place was to discover and assess the kind and degree of prejudice, you will not be wasting your time. Make sure you have clearly established your purpose in reading before giving too hasty an answer to this question.

"For whom is it being written?" This, too, can save you much time otherwise spent in reading something not intended for your attention. If you are a student of thermodynamics and you are trying to find out more about entropy, you may not want to spend your private study time reading a book intended for the interested general reader. You should always check that the book you are reading was intended for you. You cannot reasonably criticize a book for not satisfying your purpose when it did not set out to do so. This would quite clearly not be fair criticism.

How the Writer Treats His Subject

The questions about the writer's treatment of his material will give you a further guide to the writer's authority for offering you the information and opinions he does. It will also tell you more about the validity of the writer's material—that is, whether or not there is any flaw in his argument, any looseness in his expression, any serious omissions or errors of fact. You should also seek out the answers to the supplementary questions here, for these will give you the detailed picture of how the writer treats his material which you will need in order to be able to evaluate it accurately. Remember when you are doing this that excellence in style and presentation can often hide an essentially weak argument. Making such an analysis of the treatment is, therefore, rarely a waste of time.

It is also important in this connection to consider the suitability of the treatment and examine its effect upon your acceptance or rejection of what the writer says because you may find that you are not disagreeing with what the writer tells you so much as with the way he tells you. When this happens you need to make an extra effort to reach an objective assessment of the material. It is the material you should be trying to assess rather than the author, although he must figure to some extent in any judgment you make.

The Degree of the Writer's Success

If the material fails in any way to achieve what it set out to do or to fulfil the purpose for which it was intended, question 4 (a) will enable you to find out how, where, and why this is so. Thus, you will be able to place the material and value it in relation to other similar material which you have read in the past and are likely to read in the future. In criticizing constructively, it is essential to give some thought to deciding how the material could be improved if it fails to communicate effectively.

Other Points to Watch when Reading Critically

When reading critically, remember to preview the material to obtain an overview, or general impression, of the contents. Then read the material at a speed appropriate to it and to the extent of your proposed analysis. You will be able to answer some of the questions as you read and a few moments' reflection after reading will enable you to answer the others. Remember to review the material with another rapid skimming to test the validity of your conclusions and to fix important points in your mind. Finally, balance your first impressions with the detailed analysis before reaching any firm conclusions. Material may be unsatisfactory in parts but good over all and vice versa. The good critic will not condemn simply because details are imperfect, nor will he praise simply because the overall impression pleases.

When to Read Critically

These principles can be used to evaluate almost any piece of human communication, but they are more suited to certain kinds of material than to others. It is more profitable to use them where opinions are being expressed, where interpretations and recommendations are being given, and in creative work. For example, they will reveal more if applied to an important letter or report that is trying to persuade you to adopt a new method of tackling a work problem or to buy something than to an entry in a standard work of reference or a straightforward description of an industrial process.

Speed and Critical Reading

Where detailed criticism of material is necessary, there may have to be some sacrifice of speed in the interests of efficiency and accuracy, and you may have to read things several times. This is not to say that you will be wasting time, for the systematic approach outlined here will in fact save you time in the long run if properly applied. Once the material is mastered, as in study situations, reference back to the original becomes less necessary. Such documents as legal agreements and detailed technical reports will require this more careful treatment, but there will be few cases, if you are alert and attentive to your reading, in which you will not in the long run make appreciable savings in time with this approach.

You will not *necessarily* always have to read more slowly than usual to evaluate reading matter critically. Critical reading is more a question of a doubting and questioning approach than of tackling everything at a snail's pace, and you will soon find that a little practice in using the methods we have discussed will enable you to maintain reasonably high reading speeds while reading critically. In fact, on many occasions the pressure of time may not allow you to lose too much speed. You should slow down only if you have to or want to. Be flexible. Avoid complacency and a passive approach and you will make the savings in time that you require.

Other Uses of a Critical Approach

We have been primarily concerned in this chapter with the criticism of reading matter that is part of a decision-making process or part of a learning process. The critical approach can also be used in leisure reading to help you to extract greater pleasure and benefit from novels, plays, or books on hobbies and other interests. Everyone, after reading a book for pleasure, knows that he either enjoyed it or found it lacking in some way. Few people seem to be able to say why they reacted in the way they did. The use of a critical approach will help to remedy this and give your opinions a firmer basis because they have developed from a systematic examination of the material.

Some readers claim that to analyze their pleasure reading in this way decreases their enjoyment. But many others say that it increases their insight into and understanding of what they have read. The only way to find out whether it will benefit you personally is to try it out several times. Try to regard it as an attempt to increase your understanding rather than an academic exercise in literary appreciation. Be prepared for your early efforts to be disappointing and you should soon be convinced that it is worth the effort.

Other Benefits of Reading Critically

Some of the less tangible benefits of reading more critically deserve to be emphasized as encouragement to the beginner, though it is difficult to give "hard" evidence of their validity. Over a period of time, however, you will find that your critical faculties improve. You will find it becoming gradually easier and quicker for you to evaluate consciously whatever you see, hear, and read. You will find your evaluations becoming more confident and reliable and more easily communicable to others. Your understanding and appreciation of what you read will be deepened and broadened. You will find yourself reading with pleasure material which previously you would have scorned as being too esoteric or difficult. Your ability to use all the language communication skills—thinking, reading, writing, listening, and speaking—will improve steadily. Especially will this be so if, whenever you have the opportunity, you talk about your reading with others and match your own interpretations against theirs.

EXERCISES

Instructions

Preview before reading (a quick glance through). Count the time taken as part of your reading time (maximum allowed—20 seconds).

Define your purposes and expectations in reading and assess the type of writing.

Read through the passage once only *as quickly as you can without loss of comprehension.* Do not regress.

Try to improve upon your previous best performance.

Begin timing and begin reading NOW.

COLOURFUL DIVERSITY OF CULTURE
by Ruth Weiss

The people of Mauritius present a colourful diversity of cultures. Even the dry official statistics indicate this, stating that the population covers 224,076 general population (that is, whites and people of mixed African descent, the Creoles), 408,190 Hindu, 130,139 Moslem and 25,001 Chinese.

Driving through the island the visitor soon has proof of the varieties of the religion, languages and interest. In the fishing villages, on the housing estates, on sugar plantations, in the town homes, he is presented with a picture of a truly multiracial society. Hindu temples with their guardian figures, calm mosques, Christian churches, an unexpected shrine set high into a rock at a roadside indicate the mosaic of the social pattern.

The Hindu is the largest group. Indians were first imported as craftsmen, then, following the emancipation of the slaves, as indentured labourers on the sugar plantations.

Today there is still a preponderance of Hindu workers on the estates. Creoles tend to be the craftsmen and estate technicians. About a third of the sugar workers live on the plantations, the rest in near-by villages, and since the setting up of the Sugar Industry Labour Welfare Fund in 1948, their living conditions have greatly improved. The new housing estates are concrete houses, just as there is a tendency to build concrete schools. After all, Mauritius is in the cyclone belt and concrete is not as easily destroyed as flimsy shanty dwellings. There are still enough of those to be seen in villages along the roadsides.

In the villages it is usual to keep one or two cows in a shed near the house. They never leave this shelter and the whole family, particularly the women, is expected to gather the fodder in the cane fields or, in the case of a fishing village, along the beach. Milk not used in the home is sold to a middleman who peddles it to customers. He operates from a bicycle and is an important figure in the village community.

Women are also estate workers and, as is the lot of working wives anywhere, this means two jobs. The Hindu family in the rural area still lives in the traditional manner of the joint family system, which has its built-in baby-sitters and cooks. In the towns, the sophisticated way of life demands greater adaptation to the Western system, with all the resulting problems of nurseries, orphanages and old-age homes.

The Hindu community keeps to the traditional customs, beliefs and observances of Hindu religious life. Sir Seewoosagur Ramgoolam, the Prime Minister, annually takes part in the Shivarati Festival, marching part of the way with the procession of pilgrims. It is this rhythm of glamorous festivals coupled with the striking saris of the women and, of course, the abundance of oriental foods on offer in the markets and stores that have gained for Mauritius the label of "a little India beyond the seas."

The smaller Moslem group preserves its own religion and Moslems remain devout worshippers of their faith.

The least numerous of the community, the Chinese, again divide into two, one group supporting Nationalist

China, the other Mao. "How many believe what? About two thirds are Nationalists, I would think. It's almost impossible to say. The inscrutable oriental, you know," commented an Indian merchant about his Chinese friends, who in the main are the small traders, the owners of the numerous village stores.

The Creole community with its incredible range of mixtures of races, is less homogeneous than the other groups. The Creole is descended from the freed slaves, yet ironically is nearer in feeling and culture to the Franco-Mauritian than the Indo-Mauritian. The Creoles are a lively, artistic people, skilled in many different crafts; they are the fishermen and building workers of the country.

And finally, there is the white group, the Franco-Mauritian, about 2000 families numbering some 10,000 souls. They are remote and exclusive and at one time held the economic as well as the political power. Some are descended from the captains of those corsairs that disrupted British trade with the East and forced Britain to "do something about Mauritius," that is to capture the colony, which finally happened in 1810. The "Francos" are proud of their past, proud of their links with France and indeed France left its indelible mark on the island. Anyone familiar with its recent past could be forgiven for thinking he was visiting a French preserve. Officially, English is the country's language. In fact everyone speaks Creole, a French patois, while understanding both English and French and being usually more at home in the latter than the former.

Whites now mix socially with the rest on official occasions, at cocktail parties and in Church. Recently the first Mauritian Roman Catholic Bishop was consecrated, Monseigneur Jean Bargeot, and the occasion was a gathering of thousands of people from all communities and different faiths. But in their homes, whites keep themselves segregated.

Part of the public segregation is dictated by group habits. Thus the Hindu family, which could perfectly well afford an expensive suite in one of the modern hotels, prefers a beach bungalow, where many members of the family can enjoy each other's company in privacy. Other activities tend to unite the groups—everyone turns out to watch Cardiff City or the opening of the horse-racing season.

What of the colonial power? Britain still has a naval communication base on the island, which is a source of useful revenue to Mauritius. But on the whole the British are expatriates and not welded into the pattern of island society.

"Of course we feel bitter," remarked a Creole who described himself as an Anglophobe but sounded like an Anglophile. "The British never wanted us. The French even today see us as a living community, they have a dialogue with us. We are excitable, more Latin in temperament, and were considered to be nothing but of nuisance value to the British. A pity. If Britain had firmly stamped her mark on us, we might be more united. At least we would have shared a language and one culture."

Where British influence and French tradition mingle is in the legal system. The island has the Code Napoleon, while court procedure is British, a combination Mauritian lawyers applaud as getting the best of both systems. The press is as lively as the people. There are over 30 newspapers, four important dailies, and with the exception of the weekly *Mauritius Times*, they are in French.

Entertainment centres around the community life and therefore the home. Since the advent of television, clusters of people can be seen in the open air at the welfare or other centre, glued to the square box. Cinemas are popular, showing French and Indian films.

Mauritius is the nearest thing to a Welfare State; it has a good network of services and even includes a veterinary hospital.

Mauritius has been described as a racial melting pot. Only in time, however, will the ingredients truly melt. But if in the interim they co-exist peacefully, a good deal more will have been achieved than in other less complex societies. An experience during the time of the 1968 "troubles" when gang warfare broke out between Moslem and Creole youths and racial feelings were tense shows this. The leader of the Creole opposition, Gaetan Duval, who is also a barrister, was briefed to defend a Moslem charged with the murder of a Creole boy. Duval deliberately chose an almost all Creole jury. The parents of the accused broke into passionate despair, thinking the lawyer wanted to crucify their child on the altar of racial hatred. But the jury brought in a verdict of not guilty. Clearly, they were aware of the burden of responsibility placed on them. The accused was not the one on trial. The racial groups were. And the verdict of not guilty applies to them.

1300 words

Write down the time taken to read this passage and then attempt the Comprehension Test.

COMPREHENSION TEST

Select the most suitable answer in each case.
Do **not** *refer back to the passage.*

A. Retention

1. Indians were first imported to Mauritius as:
   a) slaves.
   b) building laborers.
   c) administrators.
   d) craftsmen.

2. The houses on the new housing estates are made of:
   a) wood.
   b) concrete.
   c) bricks.
   d) stone.

3. The largest population group on Mauritius is:
   a) Creole.
   b) Hindu.
   c) Moslem.
   d) Chinese.

4. Mauritius has been described as:
   a) a racial melting-pot.
   b) a miniature United Nations.
   c) a troubled society.
   d) a Chinese colony.

5. When briefed to defend a Moslem charged with the murder of a Creole boy, Gaetan Duval chose:
   a) an almost all-Moslem jury.
   b) an almost all-white jury.
   c) an almost all-Creole jury.
   d) an almost all-French jury.

B. Interpretation

6. It appears from this article that the strongest single influence upon everyday life in Mauritius is:
   a) Indian.
   b) French.
   c) Chinese.
   d) Moslem.

7. From a reading of this article one would estimate the influence of the Communist Chinese in Mauritius to be:
   a) considerable.
   b) nonexistent.
   c) declining.
   d) small.

8. The Mauritians' feelings toward the British appear to be:
   a) bitterly critical.
   b) warm and friendly.
   c) respectful.
   d) neutral.

9. Mauritian society appears to be:
   a) happily multiracial.
   b) multiracial, but with some racial problems.
   c) segregated, but with some attempts to integrate the racial groups.
   d) segregated.

10. The story at the end of the article about the Moslem who was acquitted is designed to show that:
    a) white juries are afraid to convict Moslems.
    b) lawyers are trying to crucify Creole children on the altar of racial hatred.

c) the different races in Mauritius are making serious attempts to live in peace.

d) Moslems do not kill Creoles because it is against their religion.

Convert the time taken into "words per minute" by means of the Reading Speed Conversion Table on page 254. Enter the result on the Progress Graph on page 256.

Check your answers to the Comprehension Test against the answers given on page 258. Enter the result on the Progress Graph on page 256.

C. **Discussion**—discuss one of these questions (orally if in a group, in writing if studying alone). *You may refer back to the passage.*

11. How far do you think Mauritius could, in fact, be described as a multiracial society?

12. What are the characteristics of a welfare state?

13. What is the future likely to hold for Mauritius?

14. Of what military value is the British base on Mauritius?

15. Does the rest of the world have anything to learn from the way Mauritius tackles its racial problems?

Proceed to the next reading exercise.

Instructions

Preview before reading (as before).

Define your purposes and expectations in reading and assess the type of writing.

Read through the passage **once only** *as quickly as you can without loss of comprehension.* **Do not regress.**

Try to improve upon your previous best performance.

Begin timing and begin reading NOW.

SRI LANKA'S DO-IT-YOURSELF FOOD DRIVE

by Lucien Rajakaruna

Twenty-two farmers from Sri Lanka were recently welcomed by Prime Minister Mrs. Indira Gandhi when they visited India as the guests of her Government. They were "Govi Rajas" or "Farmer Kings" from 22 districts in Sri Lanka, whose trip to India was part of the reward for achieving record yields of paddy.

The crownings of "Farmer Kings" is just one of the incentives offered by the Government in its concerted drive to make Sri Lanka self-sufficient in food. For the past three years all energies of the Government—the State-controlled radio and film unit, government and private schools, private newspapers—and all possible resources have been harnessed to help in the Food Drive. Today the Government can claim considerable success in its efforts.

The latest report of the Central Bank of Sri Lanka reveals that the Gross National Product of Sri Lanka increased by 8.3 per cent in 1968, compared with a rise of 4.4 per cent in 1967. This increase in rate of growth of the GNP is largely attributed to the success of the Government's agricultural drive.

The village cultivator now has a whole range of incentives to increased production. A free market has been created for rice with the reduction of the weekly rice subsidy to consumers, the guaranteed price for paddy has been increased, fertiliser subsidies are available and rural credit facilities have been extended. Paddy production competitions are held and the winners crowned "Farmer Kings" with attractive prizes of land, cash and even tractors.

The result of all this effort is the record crop of paddy harvested last year which was 64,600,000 bushels compared with a harvest of 55,100,000 bushels in 1967. From the time the Government launched this massive food drive in 1965 the import of rice—the country's staple diet—has declined drastically.

But the success of the Government's agricultural programme has not been confined to increased investment and
better yields. Its real success has been in its ability to enthuse
the people voluntarily to help in the national drive to attain
self-sufficiency on food. It is in this that Shramadana has
played a vital role.

The system of Shramadana, or the voluntary offering
of labour, has existed in Sri Lanka from ancient times. But
it was only in 1965, when the country was faced with a
serious crisis in food supply—due to a world shortage of
rice and dwindling foreign exchange—that the value of
Shramadana was fully realised. Shramadana, which till then
was the pet theme of social workers, religious leaders, and
a few public officers and politicians, suddenly assumed
tremendous importance. Here was a system of labour, which
if fully harnessed would nearly halve the Government's
expenditure on development work, while instilling a national
consciousness with tremendous potential for the future.

Following the Prime Minister's appeal for conscious
national effort to increase food production and the goal of
self-sufficiency, Shramadana groups were quickly organised
in most districts under the leadership of Buddhist monks,
local politicians, heads of schools, social workers and the
Government's own agricultural extension officers. A Government department set up for the purpose began directing the
activities of these groups and a much needed shot-in-the-arm
was received when the private sector, not to be left out,
offered aid in material and equipment and the World Food
Programme offered food for the volunteer workers.

These groups went into the country to restore abandoned water tanks—there are many thousand abandoned
tanks dotting the country coming down from the days of
Sinhalese kings. They undertook the cutting of irrigation
channels, the opening of sluices, assisted the traditional
cultivators in weeding their fields, transplanting paddy, the
application of fertiliser and in harvesting.

In another sphere of Shramadana, schoolchildren were
encouraged to participate in mass weeding campaigns to
assist the farmer in his most arduous task. A programme was
arranged whereby all post-primary schoolchildren would

devote one day of the year for weeding paddy fields. In the Colombo district alone, children from 1200 schools weeded 15,000 acres of paddy, contributing to at least 10 per cent of the increased yield in this region.

Following these successes, Shramadana has now been accepted as one of the most useful instruments for development, especially in the rural areas. Shramadana groups have already branched out into the field of rural housing, building of rural schools, voluntary service in hospitals, building of rural roads, clearing of swamps and the continuous flight against the weed *salvinia*, which clogs most rural waterways.

With the value of Shramadana being fully realised, the Government is now going ahead with arrangements to formulate a Shramadana Plan for the future and an island-wide Shramadana Target for each year. While the benefits of Shramadana when properly utilised are immense, the fear now is that with too much official involvement what was for centuries a traditional pattern of life may lose the basic quality of voluntary donation and soon be commercially utilised for the simple aim of reducing expenditure.

800 words

Write down the time taken to read this passage and then attempt the Comprehension Test.

COMPREHENSION TEST

Select the most suitable answer in each case.

*Do **not** refer back to the passage.*

A. Retention

1. The twenty-two farmers made the trip to India:
   a) because they were local dignitaries.
   b) as part of a reward for record yields of paddy.
   c) because they wished to protest to the prime minister.
   d) as part of a public relations campaign for Sri Lanka.

2. The increase in the rate of growth of Sri Lanka's Gross National product is attributed to:
   a) the success of the government's agricultural drive.
   b) the work of the bankers.
   c) good weather and good harvests.
   d) aid from the United States.

3. Prizes offered to farmers for increased production may even include:
   a) cars.
   b) horses.
   c) tractors.
   d) trips to the United States.

4. Shramadana is:
   a) a religious festival.
   b) the name for the Sri Lanka Parliament.
   c) money used to pay for rice.
   d) the voluntary offering of labor.

5. Schoolchildren were encouraged to participate in:
   a) driving tractors.
   b) cutting irrigation channels.
   c) mass weeding campaigns.
   d) building rural schools.

B. Interpretation

6. The food drive is being run by:
   a) the government alone.
   b) both public and individual enterprise.
   c) the people of Sri Lanka themselves.
   d) schoolchildren.

7. The food drive has succeeded because:
   a) the government has invested a great deal of money in it.
   b) it has been based on traditional habits of mutual help.
   c) the import of rice has declined drastically.
   d) the population of Sri Lanka is declining.

8. Sri Lanka produces:
   a) more rice than it needs.
   b) as much rice as it needs.

    c) less rice than it needs.

    d) none of its own rice at all.

9. The goal of the food drive is:
   a) to produce more food than Sri Lanka needs.
   b) to halve imports of rice.
   c) to double the standard of living of the people of Sri Lanka.
   d) to produce as much food as Sri Lanka needs.

10. The government's plans for the future are:
    a) being welcomed uncritically.
    b) being received with doubts in some quarters.
    c) being strongly opposed by the farmers.
    d) being revised and cut back.

Convert the time taken into "words per minute" by means of the Reading Speed Conversion Table on page 254. *Enter the result on the Progress Graph on page* 256.

Check your answers to the Comprehension Test against the answers given on page 258. *Enter the result on the Progress Graph on page* 256.

C. Discussion—discuss one of these questions (orally if in a group, in writing if studying alone). *You may refer back to the passage.*

11. Do you think the food drive will succeed?

12. Could Shramadana be effectively used as a solution in your own country's economic problems?

13. Should schoolchildren take part in national schemes like the one described in the passage or should they continue their education uninterrupted?

14. What kinds of voluntary service should young people take part in?

15. What is your own attitude to voluntary service, especially voluntary service overseas?

Proceed to the next reading exercise.

Instructions (Note: this passage is too short to preview.)

Read through the passage **once only** *as quickly as you can without loss of comprehension.* **Do not regress.**

Try to improve upon your previous best performance.

Begin timing and begin reading NOW.

IT BEGAN WITH THE WOMEN—II

by Edward Crankshaw

Next day the streets were full as they had not been since 1905. Nobody was working. There was a spontaneous general strike.

Now revolutionary politicians, caught napping the day before, came forward in an attempt to organise and exploit the situation. None of these forerunners was ever heard of again, but they served their purpose. What made it different from 1905 was that the Cossacks refused to charge the crowd: they rode about harmlessly in little groups and the demonstrators were angry, not pleading.

The Petrograd authorities did practically nothing. Perhaps they were shocked; perhaps they decided the hour to hit back had not come; perhaps they hoped the Tsar might be frightened into granting a constitution: we shall never know. Next day they arrested some of the revolutionary politicians and told the strikers to be back at work in three days.

But the following day, 11 March, the crowds were bigger than ever, and the troops were ordered to fire. A company of the Volhynian Guards defied authority by firing into the air; but later, that afternoon, they obeyed, and killed 60.

It was not just one crowd, however. There were innumerable crowds, and now they were looting and burning.

Nicholas, far away at Mogilev, ordered the Duma, the Parliament, to dissolve itself. His law, expressed through the Petrograd garrison, was the only law. But the Duma barricaded itself inside the Tauride Palace and refused to disband; and the garrison was on the edge of wholesale mutiny.

By 12 March the soldiers were making the running and the revolt had become violent. The arsenal was raided for arms; the prisoners in the dungeons of the fortress of Peter and Paul were turned loose; the headquarters of the secret police, the Okhrana, was sacked.

There was still no leadership. With little bloodshed but a great deal of noise, the people had taken over Petrograd, and now they wanted to be told what to do. The mob besieged the Tauride Palace demanding action and instructions from the Duma, which did not want to act: passive resistance to the autocracy was one thing; taking charge of the country quite another. And, anyway, what was the mood of the country? The Duma knew nothing of what was going on in the country as a whole. The Tsar was still Tsar.

But after much hesitation the Progressive Bloc of liberal politicians, under Milyukov, leader of the liberal KADETS, set up a temporary Committee of the Duma, which was to be the first Government of the new Russia, and went on sitting in the Tauride Palace.

430 words

Write down the time taken to read this passage and then attempt the Comprehension Test.

COMPREHENSION TEST

Do **not** *refer back to the passage.*

A. **Retention**

1. Who attempted to organize and exploit the situation on the second day?

2. What made the situation different from the previous rebellion in 1905?

3. What action did the Petrograd authorities take on the second day?

4. What did Nicholas order the Duma to do?

5. What was the nature of the revolt by 12 March?

B. Interpretation

6. Why was the Tsar unable to take effective action to quell the revolt?

7. Why, when they had taken over Petrograd, did the people want to be told what to do?

8. Why was the Duma reluctant to act?

9. Why did the Duma know nothing of the mood of the country as a whole?

10. Why did the crowd release prisoners from the fortress of Peter and Paul and sack the headquarters of the Okhrana?

Convert the time taken into "words per minute" by means of the Reading Speed Conversion Table on page 254. Enter the result on the Progress Graph on page 256.

Check your answers to the Comprehension Test against the answers given on page 258. Enter the result on the Progress Graph on page 256.

C. Discussion—discuss one of these questions (orally if in a group, in writing if studying alone). *You may refer back to the passage.*

11. What was the traditional relationship between the Russian people and the Cossacks?

12. Why did the 1905 rebellion fail?

13. What is the primary function of government? Why was the Duma doomed to be ineffective?

14. "Passive resistance to the autocracy was one thing; taking charge of the country quite another." Why?

15. What action should the Duma have taken at this stage of the revolt?

PRACTICE FOR THE COMING WEEK

Select any novel, read it, and then write out a concise criticism of it, using the principles and questions discussed in this chapter as a guide.

When you have done this, ask yourself if criticizing the book in this way increased or decreased:

a) your enjoyment of it.

b) your understanding of it.

Avoid regressions in reading.
Continue the practice suggested in previous chapters.

CHAPTER SUMMARY

1. **Defining Purposes in Reading**
   a) Before you read a piece of reading matter, make a brief list of your reasons for reading it.
   b) Answer the questionnaire on page 67 for a more detailed assessment of purpose.

2. **How Written Materials Are Constructed**
   a) Four types of writing to identify when reading:
      i) Description.
      ii) Exposition.
      iii) Argument.
      iv) Narrative.
   b) The paragraph is the most useful unit of expression for the efficient reader. You should try to pick out the "key" idea in each paragraph.

  c) Look for the pattern of organization in each piece of reading matter.

3. Critical Reading
  a) Consider the informational content.
  b) Consider the writer's intentions.
  c) Consider how the writer treats his subject.
  d) Consider the degree of the writer's success.

Refer to page 74 for questions to ask in putting these principles into practice.

FOUR

Flexibility and Techniques

LEVELS OF DIFFICULTY

No one would suggest for a moment that, in learning to become faster and better readers, we should allow ourselves to fall into the habit of reading everything we encounter at the same high speed. We must take account not only of our purposes in reading but also of the nature and level of difficulty of the material itself.

There are several factors which determine the level of difficulty of any particular piece of material for any individual reader and we must be aware of them if we are to approach our reading in a flexible and efficient manner.

First, there is the effect of **vocabulary**. If the vocabulary used by a writer is wide and varied or if he uses many highly specialized or technical terms, this can make the material more difficult to read. The possession of an extensive vocabulary is therefore essential to the really efficient reader if he is to overcome this problem.

Some subjects are inherently more difficult to understand than others, especially those concerned with abstract ideas, so the **subject matter** must be taken into account. Similarly, some subjects are inherently more interesting than others, especially those dealing with human life and experience, like true-life adventure stories and biographies of famous people. The **interest value of the material**, then, is the third factor to consider.

As we have already said, **our purpose in reading the material** can contribute toward making things easier or more difficult to read. There are comparatively few purposes for which perfect comprehension is essential, but where it is required this makes reading more difficult. Moreover, what you hope to obtain from the material should coincide with what the material is offering, otherwise difficulties will arise.

The **construction of material** is important. A writer must at least be competent in expressing what he has to say, otherwise he creates difficulties for the reader by, for example, poorly organized material or insufficient care in the choice of words. To a significant extent, efficient reading depends upon effective writing and this is particularly true with difficult subjects.

The **layout** of material effects the efficiency with which we can read it and here printers and publishers have a responsibility to fulfil. Typographical design, the length of the printed line, the quality and the color of the paper can all raise or lower the level of difficulty for the reader. Duplicated material can be especially difficult to read because information is frequently crammed on to a sheet of paper in an attempt at economy.

Internal and external distractions at the time of reading, in the form of noise or people moving about, or even one's attention momentarily wandering from the task of reading, can affect concentration and thus make the material more difficult to read. Even the **individual reader's personality** can make material more difficult. Some people are naturally capable of becoming interested in almost everything, others tend to specialize, and the degree of interest on the part of the reader (as opposed to the interest arising out of the material itself) can raise or lower the level of difficulty. Wide and varied reading interests can help to overcome this problem to a large extent.

Before deciding how much time, attention, and effort a reading task requires it is therefore necessary to decide whether you expect the material to be easy, of average difficulty, or difficult to read. During the coming week you should pay particular attention to this aspect of the reading process in your efforts to select the most suitable technique for each piece of reading matter.

USE OF "GEARS" IN READING SPEEDS

Flexibility involves making a conscious and deliberate choice of the most appropriate reading technique or "gear" and, in fact, reading matter can be handled in one of four ways. It can be *studied, read slowly, read rapidly,* or *skimmed.* Whichever technique is the most appropriate on any particular occasion will depend partly upon your purpose in reading and partly upon the nature and level of difficulty of the material.

Briefly, the four "gears" in reading speeds may be described as follows:

Studying. This element of reading speed involves reading, re-reading, making notes, and giving careful consideration to the full meaning and implications of the material. Quite obviously, this takes time. It is, therefore, a technique to be reserved for those occasions when the content of the material is difficult or unfamiliar and/or the material is sufficiently important for high quality of comprehension to be required. Speeds in study reading will range from a few words per minute (where a short passage is read several times, for example) to a maximum of about 200 w.p.m.

Slow Reading. For most people who have no reading efficiency training, slow reading is normal and is carried out at speeds ranging from 150 words per minute to 300 words per minute, approximately. The efficient reader uses slow reading where the material is fairly difficult or unfamiliar and/or a higher quality of comprehension than usual is required.

Rapid Reading. The technique, or "gear," which most adult readers are able to use for most purposes after a period of training such as the one provided in this book is rapid reading. It enables average or easy material to be dealt with at a comprehension level of 70 to 80 percent, which is quite adequate for most purposes in reading. Speeds range from about 300 to about 800 w.p.m.

Skimming. This technique involves allowing the eyes to move quickly across and down the page, not reading every group of words nor even every line. Effectiveness in skimming is greatly assisted by a clear sense of purpose; by particular attention to headings, subheadings, the opening and closing sentences of paragraphs; and by *key words and phrases.* It is a suitable technique when a general outline, or "overview," of the content of the material is required (800 to 1000 w.p.m.) or when the reader is trying to locate specific facts or ideas (1000+ w.p.m.). You have already begun to practice a form of skimming in your **previewing of material** before you read it and in your **reviewing of material** after having read it (*Practice for Chapter 2*).

Since you will already be quite familiar with the technique of slow reading and since you are gradually developing the ability to read rapidly in your progress through this book, let us take a closer look at the techniques of skimming and studying.

A CLOSER LOOK AT SKIMMING

Skimming, then, is a form of very fast "reading." *Your eyes move quickly across and down the page, seeking out the* **important** *information that the writer is trying to communicate and discarding everything that is of secondary or minor importance.* It can be used profitably when you are reading for specific details (we all skim when "reading" a dictionary or a telephone directory) or are looking for certain *key words and phrases* (for example, when skimming a journal article to ascertain the level at which the writer is treating his subject).

Skimming, therefore, is not reading in any normally accepted sense, but it is still a valid reading technique that can be used to obtain information from the printed page. Skilled skimmers find that they have to read in full only difficult or important material, and everything else can be skimmed without appreciable loss in comprehension. You will need to practice a great deal, however, if you are to emulate these natural skimmers. With practice, students have been known to skim at 2000 w.p.m. and even 3000 w.p.m. and still achieve high scores on the kind of comprehension test in this book.

As you practice skimming, *look* for the main points the writer is making or for the answers to certain questions about the material which may be at the back of your mind. (What is he saying that is useful to me? What evidence does he give to support his statements?) *Remember that you are here transferring the techniques that must be used to find information in dictionaries, telephone directories, handbooks, and encyclopedias to other types of material to acquire certain information only.*

When you are skimming, make the fullest possible use of the headings and subheadings provided and be particularly aware of key or "topic" sentences in paragraphs. When you have the information you need, move on quickly. Pace yourself against the clock, just as you have been doing on the reading exercises in this book. Check your comprehension by noting down main points and check these afterwards against the material. If you require a more specific comprehension test, then practice skimming on some of the earlier passages in this book that you do not recall so clearly now. You can use newspapers, journals, reports, and books for additional practice. You will also find some skimming exercises at the end of this chapter.

As you develop your skill in skimming you will find that you are benefiting considerably from the other material which the savings in time enables you to read and from not having your attention occupied by unimportant or irrelevant material. At first you may notice some apparent loss in comprehension, but, with a few days' practice, this will correct itself. After all, you are very probably skimming already when you read your daily newspaper (or do you read it *all*, even the advertisements?) All you are being asked to do here is to realize that there are many more occasions on which you can skim, and, providing you really know your purpose in reading, still understand the material well.

A CLOSER LOOK AT STUDYING

We have already defined studying briefly, but since it is a fundamental reading technique and since most people are unable to study effectively without instructions we shall now discuss it a little more fully.

As we have already said, *studying involves reading and re-reading material, making notes, and giving careful consideration to the full meaning and implications of the material.* It is important to remember that, while this process takes time and no little effort, it usually results, in the long run, in time being saved. If material is sufficiently important or difficult to require studying, any attempt to find a shortcut by being content simply to read the material will invariably result in the necessity for reading it again at a later date. If the material has been studied properly and methodically in the first place, the most you should normally have to do is to refer to your notes. We all recognize this fact when we are studying for a test, and it is surprising, therefore, that we should fail to recognize that some of our reading at work poses us similar problems. After all, if a report, for example, is to be analyzed and evaluated at a meeting, this is very similar to a test situation. On many such occasions it is we, as well as the material, who are being tested on our abilities to reach effective decisions and conclusions by our colleagues or our superiors. We cannot afford to treat such a situation lightly.

To read effectively for study purposes, you should first of all skim through the material to obtain an "overview," or general idea of the contents. *Second, you should define your purposes in studying and your expectations from the material* as clearly as possible. You will be familiar with the nature of both these steps from your work on previous chapters in this book.

The third step is to read the material at an appropriate speed, marking important words, phrases, sentences, and paragraphs or noting their substance in a notebook. This is the longest, and to many readers the most irksome, stage of the study process, but it is inevitable if the material is very important or difficult to understand. It is this stage which most readers are tempted to omit in the belief that they can obtain a high level of comprehension without it. Or it may be left out because of laziness. Whatever the reason, its omission invariably leads to having to read the material in full a second time or to suffering the consequences of lack of comprehension of the material.

The fourth step is to check your notes against the material to see that nothing of importance has been missed and to make sure that your notes provide an accurate and sufficient summary of the material. This check is most important if you are to have the confidence that when reference is made to the material in future, you need only, on most occasions, refer to your notes and do not have to re-read the material all over again. It is important that each step in the study process is completed conscientiously if the maximum benefit is to be derived from this reading technique.

On many occasions and for many purposes in reading, there will be three further steps that must be incorporated in the processes of studying. *Frequently, you will need to revise the notes you have made or the passages you have marked for special attention to ensure that your grasp of the material remains of a high order.* If this is so, you will find that the need for revision is reduced if you get into the habit of revising your notes the same day you made them. If you are studying in the evening, revise your notes early the next day. You will find that this kind of immediate revision is an effective method of fixing the material more permanently in your mind.

You will also be well-advised to attempt to relate what you have read and learned to its context in the problem being tackled, the work being done, or the subject being studied and to what has been learned from previous reading and experience. Try to see connections wherever they exist and in this way you will build up mastery of the subjects you study and establish a firm foundation for your future reading.

With the most important and the most difficult material you have to deal with, you may find it necessary to add a final step in the form of regular revision of your notes. This is particularly true if you are studying material for, say, a high-level policy meeting or some kind of test. For complete mastery you need to follow revision done immediately after reading and your attempts to relate what you have read to your previous knowledge with continuous and progressive revision until you are thoroughly familiar with the material. Regular revision is another aspect of study reading that is frequently neglected, and particular attention should be given to this. You will find that if you have followed all the previous steps in the study process then revision becomes much easier and is carried out quicker. Revision only becomes difficult and time consuming when you have *not* carried out the other steps in the process.

HANDLING WRITTEN MATERIALS SYSTEMATICALLY

One of the basic inefficiencies of the slower reader is that he tends to read slowly all the time. *Flexibility is the key to efficiency in reading.* Not every piece of writing is of equal importance and some reading requires much more time, care, and effort. *You should reserve your energies for more demanding material.* To encourage flexibility and the intelligent use of "gears," ask yourself the following questions:

1. Am I spending enough, *or too much*, time reading this material?

2. Am I taking enough, *or too much*, care over my reading on this occasion?

3. Am I making enough, *or too much*, effort to understand what I am reading?

As a general principle, the efficient reader will read everything as quickly as his purpose, the material, and conditions permit. Thus he is assured that at any time he is reading as efficiently as he is capable of reading. To achieve the kind of systematic flexibility required, you should also ask yourself, every time you read:

4. Am I reading as quickly as my purpose, the material, and conditions permit?

5. Is there anything I should be doing in order to read more efficiently?

6. Am I ready to speed up or slow down if the material suddenly becomes easier or more difficult or if my purpose in reading it changes?

You should never be "just reading," except when you are using reading merely as a time-filler.

By being flexible, you improve your ability to handle written materials systematically and to make the best possible use of the reading techniques or "gears" available to you. The following table gives an approximate guide to the kind of reading situation in which each of the four "gears" is usually used by efficient readers.

Purpose	*Material*		
	Easy	*Average*	*Difficult*
Outline only	4	4	3
General Understanding	4	3	2
Detailed Understanding	3	2	1

Key: 1 = Studying
2 = Slow reading
3 = Rapid reading
4 = Skimming

Of course, *you* are the only person who can make the final decision about how to deal with a particular piece of reading matter. You may find that, for most purposes in reading, the **preview-read-review** approach is quite adequate. If you have been practicing it, as was recommended at the end of Chapter 2, you should now be developing some skill in using this approach.

DEVELOPING FLEXIBILITY

The passages that now follow will provide you with greater opportunities than previous exercises for developing your flexibility in approaching reading materials and for doing this systematically along the lines we have discussed in this chapter and in the previous one.

The first two passages require you to summarize the content, and this should enable you to avoid trying to "spot" possible comprehension questions, leaving you free to put the principles of flexible, efficient reading into practice. Incidentally, if you have been tempted to try to "spot" questions, this break from the usual pattern of testing should help you to abandon altogether a practice which at best is of uncertain value and which at worst can cause a complete failure to understand the material.

The third passage is another one from *It Began with the Women* by Edward Crankshaw and you will be able to compare your performance on it with the two previous passages from this series.

The last four passages are designed primarily for skimming practice. Some of the passages in this chapter can also be used for practice in study reading, especially by groups that are following the course in schools, colleges, and businesses.

EXERCISES

Instructions

This passage may be used in three different ways. *You* decide how to read it.

1. **If you use the passage as a rapid reading exercise:**
   Follow the usual procedure. As a comprehension test, attempt the summary or one of the discussion questions listed at the end of the passage.

2. **If you use the passage as a skimming exercise:**
   Skim through the Declaration as quickly as you can and find the answers to as many of the following questions as possible. Time the exercise and limit yourself to **two minutes** or less. *Do not count the time it takes you to write your answers as part of your skimming time.*

a. Which article states: "Everyone has the right of life, liberty and security of person"?

b. What does Article 24 state?

c. Which article states: "Everyone, without any discrimination, has the right to equal pay for equal work"?

d. How many articles are there in the Declaration?

e. Which article states: "All human beings are born free and equal in dignity and rights"?

3. **If you use the passage as a study exercise:**
Study the Declaration carefully and then summarize its contents as instructed in the Comprehension Test at the end of the passage or attempt an answer to one of the questions for Discussion which follow it. *You may look briefly at the Comprehension Test before beginning the exercise.* A higher standard of answer should be expected than if you attempt the rapid reading exercise on this passage.

UNIVERSAL DECLARATION OF HUMAN RIGHTS

Preamble

Whereas recognition of the inherent dignity and of the equal and inalienable rights of all members of the human family is the foundation of freedom, justice and peace in the world,

Whereas disregard and contempt for human rights have resulted in barbarous acts which have outraged the conscience of mankind, and the advent of a world in which human beings shall enjoy freedom of speech and belief and freedom from fear and want has been proclaimed as the highest aspiration of the common people,

Whereas it is essential, if man is not to be compelled to have recourse, as a last resort, to rebellion against tyranny and oppression, that human rights should be protected by the rule of law,

Whereas it is essential to promote the development of friendly relations between nations,

Whereas the peoples of the United Nations have in the Charter reaffirmed their faith in fundamental human rights, in the dignity and worth of the human person and in the equal rights of men and women and have determined to promote social progress and better standards of life in larger freedom,

Whereas Member States have pledged themselves to achieve, in co-operation with the United Nations, the promotion of universal respect for and observance of human rights and fundamental freedoms,

Whereas a common understanding of these rights and freedoms is of the greatest importance for the full realisation of this pledge.

Now therefore

THE GENERAL ASSEMBLY proclaims

This Universal Declaration of Human Rights as a common standard of achievement for all peoples and all nations, to the end that every individual and every organ of society, keeping this Declaration constantly in mind, shall strive by teaching and education to promote respect for these rights and freedoms and by progressive measures, national and international, to secure their universal and effective recognition and observance, both among the peoples of Member States themselves and among the peoples of territories under their jurisdiction.

Article 1. All human beings are born free and equal in dignity and rights. They are endowed with reason and conscience and should act towards one another in a spirit of brotherhood.

Article 2. Everyone is entitled to all the rights and freedoms set forth in this Declaration, without distinction of any kind, such as race, colour, sex, language, religion, political or other opinion, national or social origin, property, birth or other status. Furthermore no distinction shall be made on the basis of the political, jurisdictional or international status of the country or territory to which a person belongs, whether it be independent, trust, non-self-governing or under any other limitations of sovereignty.

Article 3. Everyone has the right to life, liberty and security of person.

Article 4. No one shall be held in slavery or servitude; slavery and the slave trade shall be prohibited in all their forms.

Article 5. No one shall be subjected to torture or to cruel, inhuman or degrading treatment or punishment.

Article 6. Everyone has the right to recognition everywhere as a person before the law.

Article 7. All are equal before the law and are entitled without any discrimination to equal protection of the law. All are entitled to equal protection against any discrimination in violation of this Declaration and against any incitement to such discrimination.

Article 8. Everyone has the right to an effective remedy by the competent national tribunals for acts violating the fundamental rights granted him by the constitution or by law.

Article 9. No one shall be subjected to arbitrary arrest, detention or exile.

Article 10. Everyone is entitled in full equality to a fair and public hearing by an independent and impartial tribunal, in the determination of his rights and obligations and of any criminal charge against him.

Article 11.

1. Everyone charged with a penal offence has the right to be presumed innocent until proved guilty according to law in a public trial at which he has had all the guarantees necessary for his defence.

2. No one shall be held guilty of any penal offence on account of any act or omission which did not constitute a penal offence, under national or international law, at the time when it was committed. Nor shall a heavier

penalty be imposed than the one that was applicable at the time the penal offence was committed.

Article 12. No one shall be subjected to arbitrary interference with his privacy, family, home or correspondence, nor to attacks upon his honour and reputation. Everyone has the right to the protection of the law against such interference or attacks.

Article 13.

1. Everyone has the right to freedom of movement and residence within the borders of each state.
2. Everyone has the right to leave any country, including his own, and to return to his country.

Article 14.

1. Everyone has the right to seek and to enjoy in other countries asylum from persecution.
2. This right may not be invoked in the case of prosecutions genuinely arising from non-political crimes or from acts contrary to the purposes and principles of the United Nations.

Article 15.

1. Everyone has the right to a nationality.
2. No one shall be arbitrarily deprived of his nationality nor denied the right to change his nationality.

Article 16.

1. Men and women of full age, without any limitations due to race, nationality or religion, have the right to marry and to found a family. They are entitled to equal rights as to marriage and at its dissolution.
2. Marriage shall be entered into only with the free and full consent of the intending spouses.
3. The family is the natural and fundamental group unit of society and is entitled to protection by society and the State.

Article 17.

1. Everyone has the right to own property alone as well as in association with others.
2. No one shall be arbitrarily deprived of his property.

Article 18. Everyone has the right to freedom of thought, conscience and religion; this right includes freedom to change his religion or belief, and freedom, either alone or in community with others and in public or private, to manifest his religion or belief in teaching, practice, worship and observance.

Article 19. Everyone has the right to freedom of opinion and expression; this right includes freedom to hold opinions without interference and to seek, receive and impart information and ideas through any media and regardless of frontiers.

Article 20.

1. Everyone has the right to freedom of peaceful assembly and association.
2. No one may be compelled to belong to an association.

Article 21.

1. Everyone has the right to take part in the government of his country, directly or through freely chosen representatives.
2. Everyone has the right of equal access to public service in his country.
3. The will of the people shall be the basis of the authority of government; this will shall be expressed in periodic and genuine elections which shall be by universal and equal suffrage and shall be held by secret vote or by equivalent free voting procedure.

Article 22. Everyone, as a member of society, has the right to social security and is entitled to realisation, through national effort and international co-operation and in accordance with the organisation and resources of each State, of the economic, social and cultural rights indispensable for his dignity and the free development of his personality.

Article 23.

1. Everyone has the right to work, to free choice of employment, to just and favourable conditions of work and to protection against unemployment.
2. Everyone, without any discrimination, has the right to equal pay for equal work.
3. Everyone who works has the right to just and favourable remuneration insuring for himself and his family an existence worthy of human dignity, and supplemented, if necessary, by other means of social protection.
4. Everyone has the right to form and to join trade unions for the protection of his interests.

Article 24. Everyone has the right to rest and leisure, including reasonable limitation of working hours and periodic holidays with pay.

Article 25.

1. Everyone has the right to a standard of living adequate for the health and well-being of himself and of his family, including food, clothing, housing and medical care and necessary social services, and the right to security in the event of unemployment, sickness, disability, widowhood, old age or other lack of livelihood in circumstances beyond his control.
2. Motherhood and childhood are entitled to special care and assistance. All children, whether born in or out of wedlock, shall enjoy the same social protection.

Article 26.

1. Everyone has the right to education. Education shall be free at least in the elementary and fundamental stages. Elementary education shall be compulsory. Technical and Professional education shall be made generally available and higher education shall be equally accessible to all on the basis of merit.
2. Education shall be directed to the full development of the human personality and to the strengthening of respect for human rights and fundamental freedoms.

It shall promote understanding, tolerance and friendship among all nations, racial or religious groups, and shall further the activities of the United Nations for the maintenance of peace.

3. Parents have a prior right to choose the kind of education that shall be given to their children.

Article 27.

1. Everyone has the right freely to participate in the cultural life of the community, to enjoy the arts and to share in scientific advancement and its benefits.

2. Everyone has the right to the protection of the moral and material interests resulting from any scientific, literary or artistic production of which he is the author.

Article 28. Everyone is entitled to a social and international order in which the rights and freedoms set forth in this Declaration can be fully realised.

Article 29.

1. Everyone has duties to the community in which alone the free and full development of his personality is possible.

2. In the exercise of his rights and freedoms, everyone shall be subject only to such limitations as are determined by law solely for the purpose of securing due recognition and respect for the rights and freedoms of others and of meeting the just requirements of morality, public order and the general welfare in a democratic society.

3. These rights and freedoms may in no case be exercised contrary to the purposes and principles of the United Nations.

Article 30. Nothing in this Declaration may be interpreted as implying for any State, group or person any right to engage in any activity or to perform any act aimed at the destruction of any of the rights and freedoms set forth herein.

1600 words

COMPREHENSION TEST

Summary

In about 250 of your own words and *without referring back to the passage,* summarize the principal provisions of the Universal Declaration of Human Rights.

Discussion—discuss at least one of these questions (orally if in a group, in writing if studying alone). *You may refer back to the passage.*

1. How far does the Declaration set down what you would consider to be a realistic definition of human rights?

2. Which countries would you say came closest to practicing the pledges of the Declaration? Which rights have they so far failed to provide?

3. Which countries would you say have so far largely failed to grant their citizens a substantial proportion of the rights laid down in the Declaration? Of which rights are their people most in need?

4. Are there any "human rights" described in the Declaration with which you do not agree?

5. Are there any "human rights" omitted from the Declaration which you would add?

Proceed to the next reading exercise.

Instructions

This passage represents more than a "sprinting" exercise as it is longer than most of the other exercises in this book. Read it systematically, following the **preview-read-review** *approach that* *has been outlined previously. Read it as quickly as you can, within the limits set by your purposes in reading and the nature and level of difficulty of the material, as determined when you* **preview** *the passage.*

Make a note of your reading speed, but do not enter it on the progress graph, because your performance on this passage cannot really be compared with your performance on other exercises.

MARCUSE AND REVOLUTION

by Alasdair Clayre

How did Marcuse come to be a hero of the student left, in the United States, Britain and Europe? A German Professor who emigrated to America in the 1930's, who became known for his studies of Hegel, Soviet Marxism, and Freud, suddenly at the age of nearly 70—with a book called *One Dimensional Man*—seemed to be in the forefront of a world revolutionary movement. The Foreign Office has gone to the length of composing a document on Marcuse's views to circulate to our embassies.

Marcuse is one of that generation of émigrés who are still deeply affecting the arts and ideas of America, the generation that left Germany because of the rise of Hitler. He was born in Berlin in 1898; and under the Weimar republic he was an associate of the Frankfurt Institute for Social Research. During the war, he worked in the American Office of Strategic Studies, and afterwards researched at several universities in America before settling at the San Diego campus of the University of California, as Professor of Political Thought.

He is a man for whom universities have been, in his own phrase, "oases": he has exchanged one threatening environment in Germany for another 30 years later in California, and the university itself has been his defence. Last summer his telephone and electricity were cut off by forged letters, while a third letter threatened his life with a bomb. His students surrounded his house with their cars to protect him, until he left for Europe, where he continued to discuss the need for revolution, in East and West alike.

Marcuse regrets that only the students, and not the working class, have listened. "I cannot imagine a revolution without the working class," he says. Because these two forces are separated, he does not believe revolution even remotely possible, at least in America now. Nevertheless, he would allot the students a central role in its eventual preparation. Although they cannot make a revolution, they can articulate the unspoken desires of the mass of people ("unspoken," his critics would say, because they are not in fact the people's desires; "unspoken," in Marcuse's view, because the people have not yet the concepts or the freedom to express them).

The left-wing students, however, have not been content with this role alone. They have wanted to make their own revolution, and in their own immediate world, the university. A split opened between Marcuse and the leaders of the SDS (Students for a Democratic Society) at Columbia last summer. Marcuse appeared at the Theatre for Ideas in New York in the wake of the Columbia strike, to debate the possibilities of revolution in American society, with Norman Mailer and Arthur Schlesinger. At the end of the evening, he was asked what these ideas meant to him in the context of the university itself. He surprised many people in the audience by stating that he did not believe in the destruction of the universities and the setting up of counter-institutions, but that on the contrary the existing institutions could be changed, as far as was necessary, from within.

Since at that moment the most ardent revolutionaries among the students were saying they must "stop the universities functioning" and "tear them down," Marcuse's statement could easily have been taken as a repudiation of the SDS, and was understood as that by some in the audience. Marcuse did not intend this. He has since made it clear that he accepts the student demands within the university entirely, but believes that they can attain their aims by different tactics. Nevertheless, at Columbia as in Germany, some students last year felt they had left Marcuse behind. A point had been established beyond which he would not go; and they would.

Why does Marcuse say that revolution is necessary in the first place? He believes that a social system should be

condemned if it fails to satisfy the needs of its members
to the fullest extent that the available technology could
make possible. By that test, he argues, Western society fails:
its resources are devoted to waste, nuclear weapons, and
the creation of "false" needs by publicity, while extreme
inequality leaves the poor unaided, both within the West
and more desperately, outside.

Comparable criticisms apply to Soviet society, where
bureaucracy rather than business management enforces
equally false desires and satisfactions on the population.
The ailment—for Marcuse—is not specifically of the East or
of the West, but one of advanced industrial society itself.

Other systems in history have also failed to satisfy
human needs, and have been run in the interests of a ruling
group. But in the past there was always an exploited class
that could become conscious of its condition, and that could
liberate the society by revolution. And in past societies the
arts, religion, erotic life and the privacy of solitude have all
sustained potential opposition to the claims of the ruling
authority.

The special condition of our time, Marcuse thinks, is
that these areas have themselves been invaded by society.
The arts, for instance, become universally available, but at
the same time defused of their subversive reference to a
mysterious and better order. Solitude is invaded by mass
media. Silence becomes more rare. Marcuse suggests that
these developments—which he sees as in the interest of a
ruling class—are not coincidental, but products of a single
coherent system of oppression.

At the same time the working class, traditionally the
last opponent of the ruling system, has been brought into
conformity, or even enthusiastic support, by handouts of
consumer durables and security. There is no fundamental
dissent: "one-dimensional man" lives in one-dimensional
society.

And all this—Marcuse believes—has been achieved without serious cost to the rulers in either East or West, and
without dissolving "the system" that keeps them in their
present position. "One-dimensional man" accepts that the
milk will continue to arrive, and believes that as a necessary

corollary the bombs will continue to be manufactured, the poor will go on doing deadly monotonous labour, the profits or bureaucratic salaries will continue to be paid and the coloured peoples of the world will in general have to starve.

In America at least, Marcuse would say, these seem "the realities" of politics and economics, and anyone who sees them as only temporary sham realities is liable to be thought a destructive madman. For the milk does arrive; advanced industrial society does "deliver the goods." But they are only a fraction of the real good it could deliver.

Technology creates the possibility of a society without the struggle for life; without work in our sense, and hence without the need for a dominant group of rulers and managers, or for the repressive super-ego which Marcuse sees as their internal reflection. For Marcuse, the super-ego is more conservative than reality, and continues to forbid undeferred pleasure after technology has made it possible; while ruling groups—whose dominance Marcuse sees as depending on the scarcity of resources—continue to keep resources scarce, or desires artificially stimulated, in order to preserve their position.

What relevance have these ideas to Britain? Marcuse in the California sun sometimes seems remote from an English slum or a Welsh hill-farm, let alone from Calcutta. He does not seem to be writing about a country where there is still an urgent need for new building, for more material goods and for a great humanisation of the conditions of labour before leisure can become the predominant concern. Marcuse believes, however, that it is only a matter of time before all the Western countries, and maybe others, become like America in technological possibility; and that America— even if he has over-estimated its present capacity—could be an economy of the kind he describes within a few decades at most. Europe, and Britain, then, need no separate analysis in Marcuse's philosophy. Only in one respect does he think Britain unique: that it still has free speech.

These are not unique ideas. The historical framework is Marxian; and similar criticisms have already been made in Britain for the past dozen years by the New Left. Marcuse gives central place to the relation between inner life and society. He attacks not "aspects" of our society, but what he holds to be its form. And he believes that this form is not accidental, but maintained everywhere systematically by those in power, whose domination is disguised under the concept of administration. He calls our society "totalitarian," a realm of "total administration." And to counter it he brings in support from poetry, from the novel and from a tradition of philosophy that gives the greatest significance to concepts like "beauty," "justice" and "man."

Philosophy (with these concepts restored) has a central role in the revolution he foresees. It gives technology its priorities. It helps plan, in detail, the tasks that would now embody justice for the society, such as the feeding of the starving; and it allies itself with the outcasts—the very poor and the coloured races, the unemployed and the unemployable—to bring a new and more humane world into being. In this project, universities are vital; they are the meeting grounds for philosophers and scientists; they preserve the technology which would be necessary to implement the change; and, as surviving areas of comparatively free speech, they allow such ideas to be formulated and spoken in the middle of a hostile political and industrial order.

But the militant American students—for instance the leaders of the strike at Columbia last summer—saw their universities differently; not as oases of enlightenment from which the rest of society could be criticised, but as integral parts of that society, themselves important elements in the "military-industrial complex." At Columbia, for instance, the students protested that the university was collaborating in "defence" projects that were being used directly from day to day in the conduct of the Vietnam war. Towards the local Negro community they saw their university behaving like a careless landlord, tearing down Harlem houses to build its gymnasium apparently without local consultation.

The Columbia students pointed out that almost half the finance of the university came from Government sources. And when the radicals examined the structure of authority within the university they saw power apparently vested in a non-academic administration and ultimate appeal lying not to any academic authority but to a board of trustees which could contain politicians, bankers or industrialists. In such a structure, they asked, how could there be academic freedom? How could there be detached criticism of the "military-industrial complex" if that complex itself provided the money, controlled the administration and paid for research into specifically military projects?

The arguments of some faculty members at Columbia against this case did not—during the summer—convert the student revolutionaries. Instead, as violence grew, argument came to be ignored, or to seem a temporising between choices. Students took over the buildings. The police gathered. Faculty members, wearing white arm bands, patrolled the space between students and police and, when the police massed, formed a line of protection. They were swept away. The calling in of the police apparently confirmed the militants' theory about their university: in the end it was an authoritarian power structure operating under a thin disguise. More than any other event, the police action gained them sympathy and credence, both in the outside world and among their own fellow-students.

But moderate or even radical faculty members argued that—in spite of this evidence—the university in America normally does not function as the militants thought; that only in rare moments of crisis do trustees take command, or indeed interfere at all: and that administrators have little say in the day-to-day work of teaching and research.

Noam Chomsky, of the Massachusetts Institute of Technology—who himself faces possible prison for counselling resistance to the Vietnam war—regarded the extreme student case as fundamentally unreal. Universities were far more ramshackle than students supposed. There was nobody "at the top" giving orders or issuing directives. The reason why so many academics did research in subjects which aided the war, or the "empire," was not because of Government control but (in his view, far worse) because these professors were individually convinced that research of this kind was right. Their minds could only be changed individually and by example, Chomsky said; not by occupying buildings, which was irrelevant.

Meanwhile the pressures against a radical teacher were

weak; they might include some loss of promotion in certain universities, but a teacher who was swayed by these considerations was not to be taken seriously. Speech, writing and teaching within the university were genuinely free, in most cases, and amazingly free compared with the rest of American life (where "the spectrum of expressible thought that reaches any mass audience is fantastically narrow. I think it's comparable to, say, the Soviet Union. It happens to be a different part of the spectrum of thought that's presented, but it's comparable in its narrowness.").

Chomsky's way of thinking about university structure cuts against the belief in hidden but coherent systems of oppression, which seems as much a feature of the SDS case in Columbia as it is of Marcuse's indictment of modern industrial society. Where the students see a "power structure" operating secretly in the interests of a group, running the university as a sham with only the appearance of academic freedom, Chomsky sees individual professors making their free choices.

The students here are thinking in ways where Marcuse has preceded them; he has already suggested—for instance—hidden connections between Oxford philosophers with their dissolution of "the soul" and politicians, industrialists and bureaucrats with their "total administration." Such hidden systems of oppression leave men apparently free, but in reality conditioned; "satisfied," yet unaware of their true satisfactions; apparently concerned with justice, but in fact defending the interests of the ruling administrators who manipulate them.

For Marcuse this is the condition of most people in America. Certain extreme students applied this conception of society to the university itself. They were also thinking of a different "university" from Marcuse's. It was not merely the distance between Columbia and San Diego that separated them. The same institution can be a different place to those who steadily teach and research there, and to those who go there to study. Students may find themselves lost in a mass of anonymous contemporaries; sorted into classes allotted teachers by computer; taught by young graduate students who are themselves remote from the professors; graded according to a system which can seem designed mainly to provide more data for computers; categorised for some pre-fashioned role in a society that does not appear to have a place in their inner and spontaneous life.

The students were thinking also of a different outside world from that in which *One Dimensional Man* had been written only five years before. In the first place, they were much more conscious of themselves as a generation. This was perhaps inevitable, given the instantaneous linking of societies by television which can make the generation as natural a unit as the street once used to be.

In America during the 1960's a new radicalism has developed, in which the only charge that has to be made against a man is that he is over 30. The system is in a way symmetrical with military service: the old draft the young because they are under 30; the young condemn the old on the same basis of age. A language of protest which in the nineteenth century was used by the working class, and in the 1930's by the anti-Fascist Left, has come to be used by the young, of their elders, without dilution.

The other great change is the growth of violence. While the Vietnam war lasts, the American students see themselves being sent to die and kill in the same fields as the Vietnamese peasants, for a cause fewer and fewer of them believe in. As students they have draft deferments. If they are expelled for their protests these are taken away. Increasingly they see themselves as joint victims with the Vietnamese, just as Wilfred Owen came to see his whole generation, German or English, as an Isaac sacrificed by his father to the war.

Then in the cities last summer students were beaten by the police. Now, instead of seeing their role as patiently learning, in order later to help the oppressed through science and philosophy, an increasing number of them decided that they themselves were the oppressed. At first it seems grotesque that a rich white American student can feel in the same condition as a Vietnamese peasant or a Negro ghetto boy. But the use of violence, both in Vietnam and at home, made this identification, in the summer of 1968.

In the past year students turned from passive resistance to active fighting with the police. In this step—though not necessarily with the students in mind—Marcuse's thinking had preceded them. In a short essay entitled "Repressive Tolerance"—published in 1965, later than *One Dimensional Man*—he wrote: "I believe there is a natural right of resistance for oppressed and over-powered minorities to use extralegal means if the legal means have proved to be inadequate. Law and order are always and everywhere the law and order which protect the established hierarchy . . ." and again: "In terms of historical function, there is a difference between revolutionary and reactionary violence, between violence practised by the oppressed and by the oppressors. In terms of ethics both forms of violence are inhuman and evil—but since when is history made in accordance with ethical standards? To start applying them at the point where the oppressed rebel against the oppressors, the have-nots against the haves, is serving the cause of actual violence by weakening the protest against it."

In *One Dimensional Man*—published in 1964—it was the non-violent demonstrators who had been seen as the archetypes of protest, those "who go out into the streets, without arms, without protection, in order to ask for the most primitive civil rights." In the essay on tolerance, written only a year or two later, this kind of action is seen as itself "playing the game," keeping within the rules of tolerance which favour a repressive ruling class. Protest demonstrations "with *a priori* renunciation of counter-violence . . . serve to strengthen the administration by testifying to the existence of democratic liberties which, in reality, have changed their content and lost their effectiveness." The shift of position seems infinitesimal; but it foreshadows a real and perhaps an irreversible historical change.

Meanwhile the atmosphere of an SDS apartment in Columbia, or of the office of an underground newspaper in Greenwich Village or San Francisco, is itself like a coal-mine after a fall, waiting for an explosion; and the young girls who administer the free schools have a weary dedication, like the workers in refugee camps, expecting at any minute the advance of troops. But something besides violence has emerged from the sequence of events in America. Like other institutions, the universities have been made to ask what they are doing, what teaching is and what research means; the past year has given a new vitality to one tradition the universities represent: the tradition that everything human—including the university—is always open to question.

3200 words

COMPREHENSION TEST

Summary

In about 250 of your own words and *without referring back to the passage,* summarize the main points made in this article on Herbert Marcuse.

Discussion—discuss one of these questions (orally if in a group, in writing if studying alone). *You may refer back to the passage.*

1. What are Marcuse's ideas on revolution, as explained in this passage?

2. How far do you agree, or disagree, with Marcuse's analysis of modern industrial society?

3. What relevance do his ideas have today?

4. What do you think Marcuse means when he talks of "one-dimensional man"?

5. Are there any political circumstances in which violence and revolution may be justified?

Proceed to the next reading exercise.

Instructions

Read through the passage **once only** *as quickly as you can without loss of comprehension.*

Try to improve upon your previous best performance.

Begin timing and begin reading NOW.

IT BEGAN WITH THE WOMEN—III

by Edward Crankshaw

Suddenly, almost at once, there was a rival Government. Socialists of all complexions formed an emergency committee of their own. With no particular direction the left-wing revolutionary parties had coalesced into a so-called Council, or Soviet, of Workers' and Soldiers' Deputies, a resurrection of the famous Petersburg Soviet of 1905 in which the young Trotsky had played a leading part. But Trotsky was now in America.

The Soviet Committee was called the Executive Committee, or Ex-Com. The Socialist Revolutionaries and the Mensheviks in it were each separately much stronger than the Bolsheviks, who were dedicated to their destruction. But it is impossible to blame them for combining with the Bolsheviks. They knew that Lenin was implacably opposed to them; but Lenin was far away in Zurich.

It was only Lenin, literally only Lenin, who thought of his fellow revolutionaries as the greater enemy. So there was a honeymoon, and it was to last even when senior Bolsheviks such as Kamenev (Stalin, too, then 38) returned from Siberia. It was to last until Lenin himself appeared a month later.

Inside the Tauride Palace, the Duma committee could not rule because it had virtually no contact with the soldiers and the workers who had made the revolution. These came and went incessantly, but their focus was the other wing of the Palace, where the Ex-Com of the Soviet sat in perpetual session. They crowded in for instructions, for advice; they brought in recruits, they brought in prisoners, including Ministers of the Imperial Government.

They slept on the floors; they cooked in the passages; tramping in with muddy boots they filled the elegant floors with slush and filth. The air was rank with steam and reek from their greasy sheepskins and military greatcoats. They recognised only the Soviet, *their* Soviet, as the source of authority and like eager, argumentative children they pestered for guidance.

The Soviet was already a power: it had talked the Duma into agreeing that one in ten of the factory workers should be armed to form a militia to take over the duties of the police; it had set up two standing commissions to look after the food supply and to control the army. It had a link with the Duma committee in the person of Kerensky, a brilliant oratorical lawyer, a Socialist Revolutionary who somehow managed to achieve and keep membership of the Duma committee and the Soviet's Ex-Com, seeing himself increasingly as the man of the hour.

410 words

Write down the time taken to read this passage and then attempt the Comprehension Test.

COMPREHENSION TEST

Do **not** *refer back to the passage.*

A. Retention

1. Who had played a leading part in the Petersburg Soviet of 1905 but was now in America?

2. Name *two* of the member groups of the Ex-Com.

3. Where was Lenin at this time?

4. Where did the Ex-Com sit in perpetual session?

5. Who was the link man between the Ex-Com and the Duma?

B. Interpretation

6. What is a "Soviet"?

7. Why did the soldiers and workers who had made the revolution have virtually no contact with the Duma?

8. Why was the Ex-Com in perpetual session?

9. Why did Kerensky see himself as the man of the hour?

10. Why had the Soviet talked the Duma into agreeing to the setting up of a new police force and set up standing commissions to look after the food supply and to control the army?

Convert the time taken into "words per minute" by means of the Reading Speed Conversion Table on page 254. Enter the result on the Progress Graph on page 256.

Check your answers to the Comprehension Test against the answers given on page 259. Enter the result on the Progress Graph on page 256.

C. **Discussion**—discuss one of the questions (orally if in a group, in writing if studying alone). *You may refer back to the passage.*

11. What differences would there have been in the situation if Lenin and Trotsky had been in Russia at the beginning of the revolution?

12. Why did Lenin think of his fellow revolutionaries as "the greater enemy"?

13. Why did the Duma not make greater efforts to establish contact with the soldiers and workers who had made the revolution?

14. Why did the revolutionaries show such little regard for the elegance and splendour of the Tauride Palace?

15. What evidence is there in the passage for predicting that the Soviet would ultimately be successful?

PASSAGES FOR SKIMMING PRACTICE

Instructions

Skim through the passage as quickly as you can and find the answers to as many as possible of the questions below. Time the exercise and limit yourself to **three minutes** *or less.*

Do not count the time it takes you to write your answers as part of your skimming time.

1. What do the initials COIK stand for?
2. Give two names that children in the ninth grade are said not to be familiar with.
3. Who does the writer say is not familiar with the word "vapid"?
4. Which telephone number is given in the passage?
5. What substance did a scientist tell a plumber "eats hell out of the pipes"?

After you have skimmed through the passage, summarize the main points in not more than five lines.

CLEAR ONLY IF KNOWN

by Edgar Dale

For years I have puzzled over the poor communication of simple directions, especially those given me when traveling by car. I ask such seemingly easy questions as: Where do I turn off Route 30 for the bypass around the business district? How do I get to the planetarium? Or, is this the way to the university? The individual whom I hail for directions either replies, "I'm a stranger here myself," or gives me in kindly fashion the directions I request. He finishes by saying pleasantly, "You can't miss it."

But about half the time you do miss it. You turn at High Street instead of Ohio Street. It was six blocks to the turn, not seven. Many persons tell you to turn right when they mean left. You carefully count the indicated five stoplights before the turn and discover that your guide meant that blinkers should be counted as stoplights. Some of the directions turn out to be inaccurate. Your guide himself didn't know how to get there.

Education is always a problem of getting our bearings, of developing orientation, of discovering in what direction to go and how to get there. An inquiry into the problem of giving and receiving directions may help us discover something important about the educational process itself. Why do people give directions poorly and sometimes follow excellent directions inadequately?

First of all, people who give directions do not always understand the complexity of what they are communicating. They think it a simple matter to get to the Hayden Planetarium because it is simple for them. When someone says, "You can't miss it," he really means, "I can't miss it." He is suffering from what has been called the COIK fallacy—Clear Only If Known. It's easy to get to the place you are inquiring about if you already know how to get there.

We all suffer from the COIK fallacy. For example, during a World Series game a recording was made of a conversation between a rabid baseball fan and an Englishman seeing a baseball game for the first time.

The Englishman asked, "What is a pitcher?"

"He's the man down there pitching the ball to the catcher."

"But," said the Englishman, "all of the players pitch the ball and all of them catch the ball. There aren't just two persons who pitch and catch."

Later the Englishman asked, "How many strikes do you get before you are out?"

The baseball fan said, "Three."

"But," replied the Englishman, "that man struck at the ball five times before he was out."

These directions about baseball, when given to the uninitiated, are clear only if known. They are, in short, *COIK*.

Try the experiment sometime of handing a person a coat and asking him to explain how to put it on. He must assume that you have lived in the tropics, have never seen a coat worn or put on, and that he is to tell you verbally how to do it. For example, he may say, "Pick it up by the collar." This you cannot do, since you do not know what a *collar* is. He may tell you to put your arm in the sleeve or to button up the coat. But you can't follow these directions because you have no previous experience with either a sleeve or a button. He knows the subject-matter but he doesn't know how to teach it. He assumes that because it is clear to him it can easily be made clear to someone else.

The communication of teachers and pupils suffers from this COIK fallacy. An uninitiated person may think that the decimal system is easy to understand. It is—if you already know it. Some idea of the complexity of the decimal system can be gained by listening to an instructor explain the binary system—a system which many children now learn in addition to the decimal system. It is not easy to understand with just one verbal explanation. But when you understand it, you wonder why it seemed so hard.

A teacher once presented a group of parents of first-grade children with material from a first-grade reader which she had written out in shorthand, and asked them to read it. It was a frustrating experience. But these parents no longer thought it was such a simple matter to learn how to read. Reading, of course, is easy if you already know how to do it.

Sometimes our directions are overcomplex and introduce unnecessary elements. They do not follow the law of parsimony. Any unnecessary element mentioned when giving directions may prove to be a distraction. Think of the directions given for solving problems in arithmetic or for making a piece of furniture or for operating a camera. Have all unrelated and unnecessary items been eliminated? Every unnecessary step or statement is likely to increase the difficulty of reading and understanding the directions. There is no need to elaborate the obvious. Aristotle once said: "Don't go into more detail than the situation requires."

In giving directions it is easy to overestimate the experience of our questioner. It is hard indeed for a Philadelphian to understand that anyone doesn't know where the City Hall is. Certainly if you go down Broad Street, you can't miss it. We know where it is; why doesn't our questioner? Some major highways are poorly marked. In transferring to Route 128 in Massachusetts from Route 1 you must choose between signs marked "North Shore" and "South Shore." In short, you must be from Boston to understand them.

It is easy to overestimate the historical experience of a student. The college instructor may forget that college seniors were babies when Franklin D. Roosevelt died.

Children in the ninth grade are not familiar with John L. Lewis, Henry Kaiser, Quisling, Tojo. Events that the instructor has personally experienced have only been read or heard about by the student. The immediate knowledge of the instructor is mediated knowledge to the student.

Another frequent reason for failure in the communication of directions is that explanations are more technical than necessary. Thus a plumber once wrote to a research bureau pointing out that he had used hydrochloric acid to clean out sewer pipes and inquired whether there was any possible harm. The first written reply was as follows: "The efficacy of hydrochloric acid is indisputable, but the corrosive residue is incompatible with metallic permanence." The plumber then thanked them for this information approving his procedure. The dismayed research bureau wrote again, saying, "We cannot assume responsibility for the production of toxic and noxious residue with hydrochloric acid and suggest you use an alternative procedure." Once more the plumber thanked them for their approval. Finally, the bureau, worried about the New York sewers, called in a third scientist who wrote: "Don't use hydrochloric acid. It eats hell out of the pipes."

We are surprised to discover that many college freshmen do not know such words as *accrue, acquiesce, enigma, epitome, harbinger, hierarchy, lucrative, pernicious, fallacious,* and *coerce.* The average college senior does not know such words as *ingenuous, indigenous, venal, venial, vitiate, adumbrate, interment, vapid, accouterments, desultory.* These words aren't hard—if you already know them.

Some words are not understood; others are misunderstood. For example, a woman said that the doctor told her that she had "very close veins." A patient was puzzled as to how she could take two pills three times a day. A parent objected to her boy being called a scurvy elephant. He was called a disturbing element. A little boy ended the Pledge of Allegiance calling for liver, tea, and just fish for all.

Another difficulty in communicating directions lies in the unwillingness of a person to say that he doesn't know. Someone drives up and asks you where Oxford Road is. You realize that Oxford Road is somewhere in the vicinity and feel a sense of guilt about not even knowing the streets in your own town. So you tend to give poor directions instead of admitting that you don't know.

Sometimes we use the wrong medium for communicating our directions. We make them entirely verbal, and the person is thus required to hold them in mind until he has followed out each step in the directions. Think, for example, how hard it is to remember Hanford 6-7249 long enough to dial it after looking it up.

A crudely drawn map will often make our directions clear. Some indication of distance would also help, although many people give wrong estimates of distances in terms of miles. A chart or a graph can often give us at a glance an idea that is communicated verbally only with great difficulty.

But we must not put too much of the blame for inadequate directions on those who give them. Sometimes the persons who ask for help are also at fault. Communication, we must remember, is a two-way process.

Sometimes an individual doesn't understand directions but thinks he does. Only when he has lost his way does he realize that he wasn't careful enough to make sure that he really did understand. How often we let a speaker or instructor get by with such terms as "cognitive dissonance," "viable economy," "parameter," without asking the questions which might clear them up for us. Even apparently simple terms like "needs," "individual instruction," or "interests" hide many confusions. Our desire not to appear dumb, to be presumed "in the know," prevents us from understanding what has been said. Sometimes, too, the user of the term may not know what he is talking about.

We are often in too much of a hurry when we ask for directions. Like many tourists, we want to get to our destination quickly so that we can hurry back home. We don't bother to savor the trip or the scenery. So we impatiently rush off before our informant has really had time to catch his breath and make sure that we understand.

Similarly, we hurry through school and college subjects, getting a bird's-eye view of everything and a close-up of nothing. We aim to cover the ground when we should be uncovering it, probing for what is underneath the surface.

It is not easy to give directions for finding one's way around in a world whose values and directions are changing. Ancient landmarks have disappeared. What appears to be a lighthouse on the horizon turns out to be a mirage. But those who do have genuine expertness, those who possess tested, authoritative data, have an obligation to be clear in their explanations, in their presentation of ideas.

We must neither overestimate nor underestimate the knowledge of the inquiring traveler. We must avoid the COIK fallacy and realize that many of our communications are clear only if already known.

End of Passage   **1800 words**

Instructions

Skim through the passage as quickly as you can and find the answers to as many as possible of the questions below. Time the exercise and limit yourself to **four minutes** *or less.*

Do not count the time it takes you to write your answers as part of your skimming time.

1. Who won the "battle of the eyes" between Zoya and Vera?

2. In the passage, find an example of how gaze can serve as an aversive stimulus.

3. Do women look more and give more eye contact than men?

4. Name one researcher who has studied pupil size when people are looking at photographs.

5. Is it possible to state precisely when gaze will be interpreted favorably and when it will be interpreted unfavorably?

After you have skimmed through the passage, summarize the main points in not more than five lines.

I COULD TELL BY THE LOOK
IN HIS/HER EYES

by Chris Kleinke

The following ad appeared in a Boston underground newspaper:

> Beautiful woman, you boarded MTA at Central Sq. May 20, 1 p.m. You had gold shawl, violet jersey, long white print dress, sandals, large leather bag, soft brown curls, striking eyes and gray textbook. You looked up. I had curly brown hair, glasses, mustache, cord jacket, umbrella, blue valise. Our eyes met four times. I'd like to try it again. This time with words.

Apparently the man who placed the ad interpreted the four instances of eye contact as a sign that he and the young lady were interested in each other. We know from our own experience that we can communicate our attraction toward people by gazing at them. Those wishing to reciprocate our interest can return their gaze toward us. People who are not attracted can show this by looking away. But looking at people is not always taken as a sign of positive attraction. We have all been in the situation of looking at somebody longer than "allowed" and being confronted with the blunt question, "What are you looking at?" Some of you may have heard or used the expression, "Take a picture, it lasts longer." The way in which our gaze is interpreted depends so much on the situation and the particular individuals involved that it is difficult to give a general rule for predicting when looking at someone will be taken positively and when it will be received negatively. In hostile or competitive interactions between people, gaze seems to accentuate unpleasant feelings. In friendly and cooperative interactions between people, gaze is often found to increase positive feelings. Let us examine some of the different ways in which gazing behavior has been studied. You will be able to see how the effects of gaze have been found to vary in the different contexts and situations which were investigated.

GAZE AND DISCOMFORT

When we are uncomfortable with or dislike another person, we often avoid giving gaze. Receiving gaze from someone else can also be aversive, especially when it communicates threat or dominance. Most of you are familiar with the writings of animal researchers who have interpreted gaze as a display of aggression or threat and gaze aversion as a sign of submission. One investigator found that by staring at monkeys he could elicit behaviors of anger and attack.

A number of studies have shown that autistic and schizophrenic persons look significantly less at others than average people do. This avoidance of eye contact has been interpreted as an attempt to withdraw from interaction with other people, which to the autistic and schizophrenic person is emotionally arousing and uncomfortable.

College students have been found to look much less at an interviewer when the interview questions are about embarrassing issues than when the interview questions are about nonembarrassing issues. Participants in one experiment were either criticized or praised by an experimenter. The participants who were praised showed an increase in eye contact with the experimenter during the course of the experiment. The participants who were criticized decreased their eye contact with the experimenter.

As with animals, gaze between humans can serve as a sign of dominance or competition. Alexander Solzhenitsyn gives an insightful description of the operation of gaze in interpersonal conflict:

> Zoya's stare, ready to rebuff her, was so bold that Vera Kornilyevna realized that it would be impossible for her to prove anything and Zoya had already decided as much. Zoya's rebuff and her determination were so strong that Vera Kornilyevna couldn't understand them. She lowered her eyes.
>
> She always lowered her eyes when she was thinking unpleasant thoughts about someone.
>
> She lowered her eyes guiltily while Zoya, having won the battle, continued to test her with her straightforward gaze.
>
> Zoya had won the battle . . .

The relationship between gaze and dominance between people has also been studied by experimental psychologists. Male students were introduced to another male who gazed at them very often or very little. The students who received extended gaze from the other person saw him as being significantly more dominant than the students who received only brief gaze. In another study, male students were angered by another male and then were given an opportunity to vent their anger by giving shocks to him. The angered students gave significantly fewer shocks when the person who had angered them gazed at them steadily than when he did not gaze. The students reported that being gazed at by the person who had angered them was very discomforting and that they did not shock him so often because by not shocking him they could avoid his gaze.

An example of how gaze can serve as an aversive stimulus which people attempt to escape is shown in a recent field experiment conducted in California. Men and women drivers waiting at a red traffic light were confronted by a person who stared at them until the light turned green. Sometimes the staring person was on a motor scooter and sometimes the staring person was standing on the corner. Drivers who were stared at crossed the intersection with significantly greater speed after the light turned green than another group of drivers who were not stared at. Both male and female drivers showed the same motivation to escape from staring men as well as from staring women.

GAZE AND PHYSIOLOGICAL AROUSAL

In some situations physiological responses are influenced by gaze from another person. The galvanic skin response of men and women was found to be higher when they looked at someone who was gazing back at them than when they looked at someone who looked away. Galvanic skin response is a measure of skin resistance and how much we are perspiring. A high galvanic skin response means that the skin is damp and its resistance is low. This is often interpreted as an index of emotional arousal. In another study, college men played a competitive game with a male opponent who either gazed at them constantly or not at all. The men who received constant gaze from their opponent had significantly higher heart rates than the men receiving no gaze. Certain brain waves have also been shown to be affected by how much gaze a person receives. It isn't possible to state precisely what all of these physiological responses mean, but it was concluded in the above studies that gaze served as a stimulus to increase emotional arousal.

GAZE AND POSITIVE FEELING

We have seen that gaze from another person in competitive situations is likely to be interpreted as expressing hostility or dominance. In a friendly or relaxed social context we tend instead to use gaze as a measure of how much people like each other. The relationship between gaze and positive feeling has been studied in a variety of different ways.

Gaze in Videotapes

Two groups of investigators measured the reactions of people to various amounts of gaze in videotapes. In one study, men and women observed videotapes in which instructions for an experiment were read by a male who either never looked up at the camera or looked up twice. The men and women participants rated the experimenter in the videotapes as less formal and less nervous when he looked up than when he did not look up. Another study had actors portray the role of engaged couples in a videotaped interview. In half of the videotapes the "engaged" couples gazed at each other fairly often during the interview. In the other videotapes the couples never looked at each other. College students who viewed the videotapes and rated the couples were led to believe that the interviews were genuine and that the couples were actually engaged. Couples who gazed at each other were rated as significantly more genuine, relaxed, cooperative, intelligent, and attentive toward each other than couples who did not gaze. In addition, couples who gazed at each other were seen as liking each other more, as having better potential for a successful marriage, and were rated as better liked by the viewers than couples who did not gaze.

Liking People Who Gaze

You've heard this said in seriousness:

> I know he/she liked me. I could tell by the way he/she looked at me.

Others put it more humorously:

> I know he/she liked me. I could tell by the way I looked at him/her.

We often get an impression of how much people like us by the amount that they look at us. College men and women who arrived to participate in a psychology experiment were introduced briefly to two other people and asked privately to choose which of the two they would prefer to have as a partner during the experiment. Both of the people to whom the participants were introduced were actually confederates of the experimenter and were trained in such a way that one would maintain eye contact with the participant during the introduction and one would look away. Results showed that both men and women participants were significantly more likely to choose the person who looked at them during the introduction. The confederates took turns looking and not looking to show that it was their gaze and not their appearance which made the difference in whether or not they were chosen as a partner.

Males and females were asked to talk briefly about an interesting event in their lives to a listener who either leaned forward and gazed a lot or leaned back and rarely gazed. The males and females were more likely to look at listeners who looked at them. They also stated that they preferred the listeners who leaned forward and gazed over the listeners who leaned back and did not gaze.

A female experimenter gave a brief speech to a number of pairs of female undergraduate students. The experimenter arranged it so that during her talk she would look at one of the students 90 percent of the time and the other student 10 percent of the time. When students were questioned afterwards they stated that the student who had received the most gaze from the experimenter appeared to be liked best by her.

Females who interviewed males about their interests and hobbies were rated by the males as significantly more attentive if they gazed at a high rather than a low level. Males also gave significantly briefer answers to females who did not gaze at them, compared with females who gazed at them all of the time or half of the time.

Whether or not we like someone who gazes often depends on the context in which the gazing is done. Women students were interviewed individually by a female who either looked at them very often during the interview or hardly at all. In addition, for half of the participants the content of the interview was positive and complimentary and for the other participants it was negative and threatening. When asked to give their reactions toward the interviewer the students who had the positive interview preferred the interviewer when she gazed. Students who were in the negative interview, on the other hand, preferred the interviewer when she did not gaze. A similar study found that a male experimenter who gave a personal evaluation to men and women was liked most if he did not gaze very much. When giving an impersonal evaluation, the experimenter was evaluated most favorably if he gazed quite often.

The effects of gazing can also depend on the physical attractiveness of the person who gazes. Male and female students were introduced to each other and left alone to talk about anything they chose for fifteen minutes. The females were actually confederates in the experiment and were trained to gaze at their male partners either 90 percent of the time or 10 percent of the time. If the females were physically attractive it did not make much difference whether their gaze during the conversation was high or low. Unattractive females, on the other hand, were rated much more negatively by their partners when their gaze was low. Overall, attractive females were favored by males over unattractive females.

When We Like to Gaze

Besides judging people's liking by their gaze toward us, we can also communicate our liking for others by how much we gaze at them. It has been found that people gaze more when approaching a coat-rack and pretending it is a person whom they like rather than dislike. Female students who came into an experiment were told that they would be introduced to a female interviewer. Half of the students were given the instructions that they should try to gain friendship with the interviewer. The other students were instructed to attempt to avoid friendship with the interviewer. Students who were trying to make friends with the interviewer gazed at her significantly more than students who wanted to avoid her friendship.

In addition to gazing most at people we like, we also gaze more at people when they are attentive or polite toward us. Male students were asked to talk briefly to two other male listeners. One of the listeners gave approval with smiles and head nods while the other listener appeared interested but gave no approval. The students perceived the listener who gave approval as liking them more than the neutral listener and they gazed significantly more at the approving listener during their talk. People have been found to increase their eye contact with an interviewer who evaluates them positively and decrease their eye contact with an interviewer who evaluates them negatively. People also gaze more at interviewers whom they like than at interviewers whom they dislike.

Eye contact between dating couples who were very close was compared with eye contact between dating couples who were not so close. It turned out that the close couples looked at each other significantly more often than did the couples who were less close.

GAZE, HONESTY, AND PERSUASION

The amount of time that we look at someone when speaking is sometimes interpreted as communicating how truthful we are. The admonishment often given to children, "Look at me when you are speaking!" is familiar to all of us. When people are asked to tell a lie they look less at their audience than when they are asked to tell the truth. When people want to be persuasive they look at their audience more. People tend to judge verbal statements as more credible when they are accompanied with gaze.

A group of female college students went to airports and shopping centers and asked people to do them various favors, such as mailing a letter or lending a dime. Half of the time the females looked directly at the person when asking the favor and half of the time they looked down or off to the side. The differences were not remarkable, but in all cases people were more willing to comply with the requests of females who gazed than with those of females who looked away.

DIFFERENCES BETWEEN MEN AND WOMEN

Research has shown fairly consistently that women tend to look more and give more eye contact than men. This might be because women are generally more open to intimacy than men or because of a greater desire by women to seek feedback about how others are reacting toward them.

Male and female college students were introduced in pairs and left alone in a room to talk about anything they chose for ten minutes. After ten minutes, a male experimenter came into the room and told the couple that he had been measuring through a hidden one-way mirror how much one of them had looked at the other during the conversation. Half of the time the experimenter had supposedly measured how much the male looked at the female and half of the time the measure was supposedly of how much the female had looked at the male. The experimenter told the couple that the amount of time the one person had looked at the other was significantly longer than most people look, about average, or significantly shorter. This feedback had nothing to do with how much the people had actually looked, but was designed to influence the perceptions that the participants had of their own looking behavior or the looking behavior of their partner. The false feedback about gaze was believable to the participants because there actually was a one-way mirror in the room and because most of us aren't aware enough of how much we look at people to question the feedback of an experimenter. The point of all this discussion is that females reacted most positively when they thought they or their partner had gazed at a very high level. Males, on the other hand, were not comfortable with the idea that they might have given or received above average amounts of gaze. Males reacted most positively when they thought their gaze or the gaze of their partner was average.

PUPIL SIZE

Eckhard Hess tells the story that he and his wife were in bed reading one night when he happened to notice that his wife's pupils changed size as she came to different parts of her book. It occurred to Hess that pupil size might have something to do with one's interest or favor in whatever he or she is looking at. As a result of Hess' insight, a number of studies have been conducted showing the pupil sizes increase when people are looking at favorable or positive stimuli. Pupils of males become larger when they are looking at pictures of nude females. People looking at slides of political figures have larger pupils when the picture is of someone they like and smaller pupils when the picture is of someone they dislike. Homosexual and heterosexual males have been differentiated by showing that homosexuals have larger pupils when looking at pictures of men.

Hess also investigated whether or not pupil size would have any influence on first impressions. People were asked to give their reactions to an attractive woman in two photographs which were identical in all aspects but one. In one of the photographs the woman's pupils had been made to look slightly larger. Hess found that people preferred the woman when her pupils were large.

An interesting study of pupil size was conducted with

people in a real interaction. Participants came to take part in an experiment and were asked privately to choose one of two people as their partner. One of the people had pupils which were dilated with a drug. The other person had normal pupils. When choosing between two females, both men and women participants favored the female with enlarged pupils. When choosing between two males, both men and women participants preferred the male with enlarged pupils. Care was taken to vary which male or female stimulus people would have enlarged and normal pupils in order to be certain that it was really pupil size and not other factors, such as relative attractiveness or unattractiveness, which influenced the participants' choices.

TO GAZE OR NOT TO GAZE

You can see now why I said earlier that it is impossible to state precisely when gaze will be interpreted favorably and when it will be interpreted unfavorably. The effects of gaze depend on factors having to do with the situation and the person who is gazing which are too complex at this point to define. In some situations, and in the association of some facial expressions, gaze can be an aversive stimulus. In other contexts, gaze can be a sign of attraction and liking. It is possible that gaze serves mainly to accentuate whatever feelings are present in a given situation. In a pleasant encounter gaze might increase the pleasantness and in an unpleasant encounter gaze might function to increase feelings of discomfort. You may be interested in seeing what happens when you gaze or do not gaze at various people you encounter. Try observing other people and see when and how much they look at each other. But don't blame me if they ask you what you're looking at!

End of Passage   3300 words

Instructions

Skim through the passage as quickly as you can and find the answers to as many as possible of the questions below. Time the exercise and limit yourself to **five minutes** *or less.*

Do not count the time it takes you to write your answers as part of your skimming time.

1. How much space, according to a formula of obscure origin, does a man in a crowd require?

2. Name one of the films quoted as causing waiting cinema-goers to stand closer together.

3. Who has researched the spatial behaviors of fowl?

4. Name an ethologist referred to in the passage.

5. Give an example, quoted in the passage, of a "non-person."

After you have skimmed through the passage, summarize the main points in not more than five lines.

SPATIAL INVASION

by Robert Sommer

Dear Abby: I have a pet peeve that sounds so petty and stupid that I'm almost ashamed to mention it. It is people who come and sit down beside me on the piano bench while I'm playing. I don't know why this bothers me so much, but it does. Now you know, Abby, you can't tell someone to get up and go sit somewhere else without hurting their feelings. But it would be a big relief to me if I could get them to move in a nice inoffensive way . . .

Lost Chord

Dear Lost: People want to sit beside you while you're playing because they are fascinated. Change your attitude and regard their presence as a compliment, and it might be easier to bear. P.S. You might also change your piano bench for a piano stool. (Abigail Van Buren, *San Francisco Chronicle,* May 25, 1965)

The best way to learn the location of invisible boundaries is to keep walking until somebody complains. Personal space refers to an area with invisible boundaries surrounding a person's body into which intruders may not come. Like the porcupines in Schopenhauer's fable, people like to be close enough to obtain warmth and comradeship but far enough away to avoid pricking one another. Personal space is not necessarily spherical in shape, nor does it extend equally in all directions. (People are able to tolerate closer presence of a stranger at their sides than directly in front.) It has been likened to a snail shell, a soap bubble, an aura, and "breathing room." There are major differences between cultures in the distances that people maintain—Englishmen keep further apart then Frenchmen or South Americans. Reports from Hong Kong where three million people are crowded into 12 square miles indicate that the population has adapted to the crowding reasonably well. The Hong Kong Housing Authority, now in its tenth year of operation, builds and manages low-cost apartments for families that provide approximately 35 square feet per person for living-sleeping accommodations. When the construction supervisor of one Hong Kong project was asked what the effects of doubling the amount of floor area would be upon the living patterns, he replied, "With 60 square feet per person, the tenants would sublet!"

Although some people claim to see a characteristic aura around human bodies and are able to describe its color, luminosity, and dimensions, most observers cannot confirm these reports and must evolve a concept of personal space from interpersonal transactions. There is a considerable similarity between personal space and *individual distance,* or the characteristic spacing of species members. Individual distance exists only when two or more members of the same species are present and is greatly affected by population density and territorial behavior. Individual distance and personal space interact to affect the distribution of persons. The violation of individual distance is the violation of society's expectations; the invasion of personal space is an intrusion into a person's self-boundaries. Individual distance may be outside the area of personal space—conversation between two chairs across the room exceeds the boundaries of personal space, or individual distance may be less than the boundaries of personal space—sitting next to someone on a piano bench is within the expected distance but also within the bounds of personal space and may cause discomfort to the player. If there is only one individual present, there is infinite individual distance, which is why it is useful to maintain a concept of personal space, which has also been described as a *portable territory,* since the individual carries it with him wherever he goes although it disappears under certain conditions, such as crowding.

There is a formula of obscure origin that a man in a crowd requires at least two square feet. This is an absolute minimum and applies, according to one authority, to a thin man in a subway. A fat man would require twice as much space or more. Journalist Herbert Jacobs became interested in spatial behavior when he was a reporter covering political rallies. Jacobs found that estimates of crowd size varied with the observer's politics. Some estimates by police and politicians were shown to be twenty times larger than the crowd size derived from head count or aerial photographs. Jacobs found a fertile field for his research on the Berkeley campus where outdoor rallies are frequent throughout the year. He concluded that people in dense crowds have six to eight square feet each, while in loose crowds, with people moving in and out, there is an average of ten square feet per person. Jacobs' formula is that crowd size equals length X width of the crowd divided by the appropriate correction factor depending upon whether the crowd is dense or loose. On the Berkeley campus this produced estimates reasonably close to those obtained from aerial photographs.

Hospital patients complain not only that their personal space and their very bodies are continually violated by nurses, interns, and physicians who do not bother to introduce themselves or explain their activities, but that their territories are violated by well-meaning visitors who will ignore "No Visitors" signs. Frequently patients are too sick or too sensitive to repel intruders. Once surgery is finished or the medical treatment has been instituted, the patient is left to his own devices to find peace and privacy. John Lear, the science editor of the *Saturday Review,* noticed an interesting hospital game he called, "Never Close the Door," when he was a surgery patient. Although his physician wanted him protected against outside noises and distractions, the door opened at intervals, people peered in, sometimes entered, but no one ever closed the door. When Lear protested, he was met by hostile looks and indignant remarks such as, "I'm only trying to do my job, Mister." It was never clear to Lear why the job—whatever it was—required the intruder to leave the door ajar afterwards.

Spatial invasions are not uncommon during police interrogations. One police textbook recommends that the interrogator should sit close to the suspect, with no table or desk between them, since "an obstruction of any sort affords the subject a certain degree of relief and confidence not otherwise obtainable." At the beginning of the session, the officer's chair may be two or three feet away, "but after the interrogation is under way the interrogator should move his chair in closer so that ultimately one of the subject's knees is just about in between the interrogator's two knees."

Lovers pressed together close their eyes when they kiss. On intimate occasions the lights are typically dim to reduce not only the distracting external cues but also to permit two people to remain close together. Personal space is a culturally acquired daylight phenomenon. Strangers are affected differently than friends by a loss of personal space. During rush hour, subway riders lower their eyes and sometimes "freeze" or become rigid as a form of minimizing unwanted social intercourse. Boy-meets-girl on a crowded rush hour train would be a logical plot for an American theater based largely in New York City, but it is rarely used. The idea of meeting someone under conditions where privacy, dignity, and individuality are so reduced is difficult to accept.

A driver can make another exceedingly nervous by tailgating. Highway authorities recommend a "space cushion" of at least one car length for every ten miles per hour of speed. You can buy a bumper sticker or a lapel button with the message "If you can read this, you're too close." A perceptive suburban theater owner noticed the way crowds arranged themselves in his lobby for different pictures. His lobby was designed to hold approximately 200 customers who would wait behind a roped area for the theater to clear.

> When we play a [family picture like] *Mary Poppins, Born Free*, or *The Cardinal*, we can line up only about 100 to 125 people. These patrons stand about a foot apart and don't touch the person next to them. But when we play a [sex comedy like]

Tom Jones or *Irma la Douce,* we can get 300 to 350 in the same space. These people stand so close to each other you'd think they were all going to the same home at the end of the show!

Animal studies indicate that individual distance is learned during the early years. At some stage early in his life the individual learns how far he must stay from species members. When he is deprived of contact with his own kind, as in isolation studies, he cannot learn proper spacing, which sets him up as a failure in subsequent social intercourse—he comes too close and evokes threat displays or stays too far away to be considered a member of the group. Newborn of many species can be induced to follow novel stimuli in place of their parents. If a newly hatched chicken is separated from his mother and shown a flashing light instead, on subsequent occasions he will follow the flashing light rather than his mother. The distance he remains behind the object is a function of its size; young chicks will remain further behind a large object than a small one.

Probably the most feasible method for exploring individual distance and personal space with their invisible boundaries is to approach people and observe their reactions. Individual distance is not an absolute figure but varies with the relationship between the individuals, the distance at which others in the situation are placed, and the bodily orientations of the individuals one to another. The most systematic work along these lines has been undertaken by the anthropologist Ray Birdwhistell who records a person's response with zoom lenses and is able to detect even minute eye movements and hand tremors as the invader approaches the emotionally egotistic zone around the victim.

One of the earliest attempts to invade personal space on a systematic basis was undertaken by Williams, who wanted to learn how different people would react to excessive closeness. Classifying students as introverts or extroverts on the basis of their scores on a personality test, he placed each individual in an experimental room and then walked toward the person, telling him to speak out as soon as he (Williams) came too close. Afterward he used the reverse condition, starting at a point very close and moving away until the person reported that he was too far away for comfortable conversation. His results showed that introverts kept people at a greater conversational distance than extroverts.

The same conclusion was reached by Leipold, who studied the distance at which introverted and extroverted college students placed themselves in relation to an interviewer in either a stress or a nonstress situation. When the student entered the experimental room, he was given either the stress, praise, or neutral instructions. The stress instructions were, "We feel that your course grade is quite poor and that you have not tried your best. Please take a seat in the next room and Mr. Leipold will be in shortly to discuss this with you." The neutral control instructions read, "Mr. Leipold is interested in your feelings about the introductory course. Would you please take a seat in the next room." After the student had entered and seated himself, Mr. Leipold came in, recorded the student's seating position, and conducted the interview. The results showed that students given praise sat closest to Leipold's chair, followed by those in the neutral condition, with students given the stress instructions maintaining the most distance from Leipold's chair behind the desk. It was also found that introverted and anxious individuals sat further away from him than did extroverted students with a lower anxiety level.

Glen McBride has done some excellent work on the spatial behaviors of fowl, not only in captivity but in their feral state on islands off the Australian coast. He has recently turned his attention to human spatial behavior using the galvanic skin response (GSR) as an index of emotionality. The GSR picks up changes in skin conductivity that relate to stress and emotional behavior. The same principle underlies what is popularly known as the lie detector test. McBride placed college students in a chair from which they were approached by both male and female experimenters as well as by paper figures and nonhuman objects. It was found that GSR was greatest (skin resistance was least) when a person was approached frontally, whereas a side approach yielded a greater response than a rear approach. The students reacted more strongly to the approach of someone of the opposite sex than to someone of the same sex. Being touched by an object produced less of a GSR than being touched by a person.

A similar procedure without the GSR apparatus was used by Argyle and Dean, who invited their subjects to participate in a perceptual experiment in which they were to "stand as close as comfortable to see well" to a book, a plaster head, and a cut-out life-size photograph of the senior author with his eyes closed and other photograph with his eyes open. Among other results, it was found that the subjects placed themselves closer to the eyes-closed photograph than the eyes-open photograph. Horowitz, Duff, and Stratton used a similar procedure with schizophrenic and nonschizophrenic mental patients. Each individual was instructed to walk over to a person, or in another condition a hatrack, and the distance between the goal and his stopping place was measured. It was found that most people came closer to the hatrack than they did to another person. Each tended to have a characteristic individual distance that was relatively stable from one situation to another, but was shorter for inanimate objects than for people. Schizophrenics generally kept greater distance between themselves and others than did nonpatients. The last finding is based on average distance values, which could be somewhat inflated by a few schizophrenics who maintain a large individual distance. Another study showed that some schizophrenic patients sat "too close" and made other people nervous by doing this. However, it was more often the case that schizophrenics maintained excessive physical distance to reduce the prospects of unwanted social intercourse.

In order to explore personal space using the invasion technique, but to avoid the usual connotations surrounding forced close proximity to strangers, my own method was to undertake the invasion in a place where the usual sanctions of the outside world did not apply. Deliberate invasions of personal space seem more feasible and appropriate inside a mental hospital than outside. Afterward, it became apparent that this method could be adapted for use in other settings such as the library in which Nancy Russo spent many hours sitting too close to other girls.

The first study took place at a 1500-bed mental institution situated in parklike surroundings in northern California. Most wards were unlocked, and patients spent considerable time out of doors. In wooded areas it was common to see patients seated under the trees, one to a bench or knoll. The wards within the buildings were relatively empty during the day because of the number of patients outside as well as those who worked in hospital industry. This made it possible for patients to isolate themselves from others by finding a deserted area on the grounds or remaining in an almost empty building. At the outset I spent considerable time observing how patients isolated themselves from one another. One man typically sat at the base of a fire escape so he was protected by the bushes on one side and the railing on the other. Others would lie on benches in remote areas and feign sleep if approached. On the wards a patient might sit in a corner and place magazines or his coat on adjacent seats to protect the space. The use of belongings to indicate possession is very common in bus stations, cafeterias, waiting rooms, but the mental patient is limited in using this method since he lacks possessions. Were he to own a magazine or book, which is unlikely, and left it on an empty chair, it would quickly vanish.

Prospective victims had to meet three criteria—male, sitting alone, and not engaged in any definite activity such as reading or playing cards. When a patient fitting these criteria was located, I walked over and sat beside him without saying a word. If the patient moved his chair or slid further down the bench, I moved a like distance to keep the space between us to about six inches. In all sessions I juggled my key ring a few times to assert my dominance, the key being a mark of status in a mental hospital. It can be noted that these sessions not only invaded the patient's personal space but also the nurse's territory. It bothered the nurses to see a high status person (jacket, white shirt, tie, and the title "Doctor") entering their wards and sitting among the patients. The dayroom was the patients' territory vis-à-vis the nurses, but it was the nurses' territory vis-à-vis the medical staff. Control subjects were selected from other patients who were seated some distance away but whose actions could be observed.

Within two minutes, all of the control subjects remained but one-third of the invasion victims had been driven away. Within nine minutes, fully half of the victims had departed compared with only 8 per cent of the controls. Flight was a gross reaction to the intrusion; there were many more subtle indications of the patient's discomfort. The typical sequence was for the victim to face away immediately, pull in his shoulders, and place his elbows at his sides. Facing away was an almost universal reaction among the victims, often coupled with hands placed against the chin as a buffer. Records obtained during the notetaking sessions illustrate this defensive pattern.

Example A

10:00. Seat myself next to a patient, about sixty years of age; he is smoking and possibly watching TV.

10:04. Patient rubs his face briefly with the back of his hand.

10:05. Patient breathes heavily, still smoking, and puts his ashes into a tin can. He looks at his watch occasionally.

10:06. He puts out his cigarette, rubs his face with the back of his hand. Still watching TV.

10:12. Patient glances at his watch, flexes his fingers.

10:13. Patient rises, walks over and sits at a seat several chairs over. Observation ended.

Example B

8:46. Seat myself next to a 60-year-old man who looks up as I enter the room. As I sit down, he begins talking to himself, snuffs out his cigarette, rises, and walks across the room. He asks a patient on the other side of the room, "You want me to sit here?" He starts arranging the chairs against the wall and finally sits down.

Ethologist Ewan Grant has made a detailed analysis of the patient's micro behaviors drawing much inspiration from the work of Tinbergen as well as his own previous studies with colonies of monkeys and rats. Among a group of confined mental patients he determined a relatively straightforward dominance hierarchy based on aggression-flight encounters between individuals. Aggressive acts included threat gestures ("a direct look plus a sharp movement of the head towards the other person"), frowns, and hand-raising.

Flight behaviors included retreat, bodily evasions, closed eyes, withdrawing the chin into the chest, hunching, and crouching. These defensive behaviors occurred when a dominant individual sat too close to a subordinate. This could be preceded by some overt sign of tension such as rocking, leg swinging, or tapping. Grant describes one such encounter: "A lower ranking member of the group is sitting in a chair; a dominant approaches and sits near her. The first patient begins to rock and then frequently, on one of the forward movements, she gets up and moves away."

In seventeen British old folks' homes Lipman found that most of the patients had favorite chairs that they considered "theirs." Their title to these chairs was supported by the behavior of both patients and staff. A newly admitted inmate had great difficulty in finding a chair that was not owned by anyone. Typically he occupied one seat and then another until he found one that was "unowned." When he sat in someone else's chair, he was told to move away in no uncertain terms. Accidental invasions were an accepted fact of life in these old folks' homes. It is possible to view them as a hazing or initiation ceremony for new residents to teach them the informal institutional rules and understandings. Such situations illustrate the importance of knowing not only how people mark out and personalize spaces, but how they respond to intrusions.

We come now to the sessions Nancy Russo conducted in the study hall of a college library, a large high-ceilinged room with book-lined walls. Because this is a study area, students typically try to space themselves as far as possible from one another. Systematic observations over a two-year period disclosed that the first occupants of the room generally sat one to a table at end chairs. Her victims were all females sitting alone with at least one book in front of them and empty chairs on either side and across. In other words, the prospective victim was sitting in an area surrounded by empty chairs, which indicated something about her preference for solitude as well as making an invasion relatively easy. The second female to meet these criteria in each session and who was visible to Mrs. Russo served as a control. Each control subject was observed from a distance and no invasion was attempted. There were five different approaches used in the invasions—sometimes Mrs. Russo would sit alongside the subject, other times directly across from her, and so forth. All of these were violations of the typical seating norms in the library, which required a newcomer to sit at a considerable distance from those already seated unless the room was crowded.

Occupying the adjacent chair and moving it closer to the victim produced the quickest departures, and there was a slight but also significant difference between the other invasion locations and the control condition. There were wide individual differences in the ways the victims reacted—there is no single reaction to someone's sitting too close; there are defensive gestures, shifts in posture, and attempts to move away. If these fail or are ignored by the invader, or he shifts position too, the victim eventually takes to flight. Crook measured spacing of birds in three ways: *arrival distance* or how far from settled birds a newcomer will land, *settled distance* or the resultant distance after adjustments have occurred, and the *distance after departure* or how far apart birds remain after intermediate birds have left. The methods employed in the mental hospital and portions of the library study when the invader shifted his position if the victim moved, maintained the *arrival distance* and did not permit the victim to achieve a comfortable *settled distance*. It is noteworthy that the preponderance of flight reactions occurred under these conditions. There was a dearth of direct verbal responses to the invasions. Only two of the 69 mental patients and one of the 80 students asked the invader to move over. This provides support for Edward Hall's view that "we treat space somewhat as we treat sex. It is there, but we don't talk about it."

Architecture students on the Berkeley campus now undertake behavioral studies as part of their training. One team noted the reactions of students on outdoor benches when an experimenter joined them on the same bench. The occupant shifted position more frequently in a specified time-frame and left the bench earlier than control subjects who were alone. A second team was interested in individual distance on ten-foot benches. When the experimenter seated himself one foot from the end of the bench, three-quarters of the next occupants sat six to eight feet away, and almost half placed books or coats as barriers between themselves and the experimenter. Another two students studied eyeblink and shifts in body position as related to whether a stranger sat facing someone or sat facing away. Observations were made by a second experimenter using binoculars from a distance. A male stranger directly facing a female markedly increased her eyeblink rate as well as body movements but had no discernible effect on male subjects.

The different ways in which victims react to invasions may also be due to variations in the perception of the expected distance or in the ability to concentrate. It has been demonstrated that the individual distance between birds is reduced when one bird's attention is riveted to some activity. The invasions can also be looked at as nonverbal communication with the victims receiving messages ranging from "This girl considers me a nonperson" to "This girl is making a sexual advance." The fact that regressed and "burnt out" patients can be moved from their places by sheer propinquity is of theoretical and practical importance. In view of the difficulty that nurses and others have in obtaining any response at all from these patients, it is noteworthy that an emotion sufficient to generate flight can be produced simply by sitting alongside them. More recently we have been experimenting with visual invasions, or attempts to dislodge someone from his place by staring directly at him. In the library at least, the evil eye seems less effective than a spatial invasion since the victims are able to lose themselves in their books. If they could not escape eye contact so easily, this method might be more effective. Mrs. Russo was sensitive to her own feelings during these sessions and described how she "lost her cool" when a victim looked directly toward her. Eye contact produced a sudden realization that "this is a human being," which subsided when the victim turned away. Civil rights demonstrators attempt to preserve their human dignity by maintaining eye contact with their adversaries.

There are other sorts of invasions—auditory assaults in which strangers press personal narratives on hapless seatmates on airplanes and buses, and olfactory invasions long celebrated in television commercials. Another interesting situation is the two-person invasion. On the basis of animal work, particularly on chickens crowded in coops, it was discovered that when a subordinate encounters a dominant at very close quarters where flight is difficult or impossible, the subordinate is likely to freeze in his tracks until the dominant departs or at least looks away. Two faculty members sitting on either side of a student or two physicians on either side of a patient's bed would probably produce this type of freezing. The victim would be unlikely to move until he had some sign that the dominants had their attention elsewhere.

The library studies made clear that an important consideration in defining a spatial invasion is whether the parties involved perceive one another as persons. A nonperson cannot invade someone's personal space any more than a tree or chair can. It is common under certain conditions for one person to react to another as an object or part of the background. Examples would be the hospital nurses who discuss a patient's condition at his bedside, seemingly oblivious to his presence, the Negro maid in the white home who serves dinner while the husband and wife discuss the race question, and the janitor who enters an office without knocking to empty the wastebaskets while the occupant is making an important phone call. Many subway riders who have adjusted to crowding through psychological withdrawal prefer to treat other riders as nonpersons and keenly resent situations, such as a stop so abrupt that the person alongside pushes into them *and then apologizes,* when the other rider becomes a person. There are also riders who dislike the lonely alienated condition of subway travel and look forward to emergency situations in which people become real. When a lost child is looking for his mother, a person has been hurt, or a car is stalled down the tracks, strangers are allowed to talk to one another.

End of Passage   **4600 words**

PRACTICE FOR THE COMING WEEK

Each day skim through at least one fairly long (about 1000 words) newspaper article. Test comprehension by noting down main points and checking against the passage for corrections.

Each week from now on, skim through a book relevant to your work, studies, or some other purpose. Select a book you would not otherwise have read.

Decide on which occasions you should be study reading. Use the systematic approach outlined in this chapter. Do you get better results or not?

If you require additional practice, continue with some of the exercises suggested earlier.

CHAPTER SUMMARY

1. **Levels of Difficulty**
   Factors that determine the level of difficulty of a piece of reading material include:
   a) vocabulary.
   b) subject matter.
   c) interest value of material.
   d) purpose in reading.
   e) construction of material.
   f) layout.
   g) internal and external distractions.
   h) individual reader's personality.

2. **Use of "Gears" in Reading Speeds**
   a) There are four "gears" in reading speeds:
       i) studying.
       ii) slow reading.
       iii) rapid reading.
       iv) skimming.

b) Skimming is not reading in the normal sense, but it is a valuable reading technique in which skill increases with practice.

c) Studying should be carried out systematically.

3. **Handling Written Materials Systematically**

   a) Flexibility is the key to efficiency in reading.

   b) The efficient reader will read everything as quickly as his purpose, the material, and conditions permit.

   c) Practice on a variety of materials read for a variety of purposes encourages the development of flexibility.

FIVE

Some Problems in Reading

So far we have mainly been concerned with general problems of efficiency in reading. Now we shall consider a few problems encountered in particular kinds of reading material. Comments will be offered on five kinds of reading matter: newspapers; correspondence; journals and magazines; business, industrial and technical reports; and leisure reading (although, strictly speaking, this is outside the scope of this book, readers may welcome some comments). We shall not try to make definitive and detailed pronouncements on how these materials should be read. This is a matter that only the individual reader can resolve in the light of his purposes in reading and the precise nature of the material he is reading. However, some comments may still be helpful. They are based on problems described by students who have attended the author's reading improvement courses.

NEWSPAPERS

Daily morning newspapers can contain anything from eight pages upwards and evening newspapers, which most of us also read, can be as large. If we read more than one morning paper, the problem is greater and if we are to deal efficiently with this amount of daily reading, we have to develop methodical, time-saving approaches.

Many newspapers have a declared or a concealed political bias and it is desirable for us to read more than one if we are to obtain a balanced picture of the news. If we are unable to find time to do this, then we are perhaps wiser if we select a newspaper whose bias is obvious and we can compensate for this in our assessments of the validity of its comments on current events and of its selection of the news to be presented to us.

Another difficulty, with many newspapers, is that of finding significant news among a mass of triviality and advertising. With some newspapers, we cannot even be certain that the most important news items will appear on the front page.

Most readers, even those who are otherwise inefficient, have some kind of a method for dealing with newspapers. After we have been reading a particular newspaper for only a few days, we know approximately where to find the news pages, feature articles, sports pages, and information about the weather, radio and television programs, and so on. Since you will know your own newspaper well, you should use this knowledge as a basis upon which to develop a more efficient method. You should **skim** the page of a newspaper quickly, selecting, by previewing, the items and articles you will read. In doing this it may be useful to make a mental note of articles that begin on one page and are continued on another. You will reduce page-turning and thus increase speed of handling if you leave these articles until you have read those which begin and end on the page you are dealing with at the moment. Read the opening few paragraphs of the articles you have selected. This will enable you to decide whether the item merits a full reading. Remember also that *news articles are frequently constructed so that the most important information is contained in the opening paragraphs.* The reasons for this are that most readers will not read an article unless tempted by attractive presentation, which involves giving the essence of the story first, and that new stories may require space in later editions. Since it is easier to cut an article from the end rather than the beginning, the item can be shortened if necessary with the loss of some of the details only and not the main points. You should, of course, read newspapers critically and enable yourself to do this more effectively by making comparisons between papers whenever possible.

Many readers also find that the layout of newspapers creates problems. *Often, material is not presented in the familiar sentences and paragraphs of conventional expression.* Paragraphs are condensed into sentences and sentences into phrases. This is done, in the "popular" press, for two reasons. First, it enables those responsible for producing the newspaper to improve its appearance and make it appear easy and attractive to read. Second, it *does* make the newspaper easy to read, but *readers who rely too much on the popular press for their information will find it more difficult to handle other materials efficiently.* If you read a "popular" paper at present, try a more "serious" newspaper occasionally. Variety is as important in newspaper reading as in reading other materials.

CORRESPONDENCE

If you have a great deal of correspondence to handle every day you will probably agree that one problem with correspondence is that few people really sit down and consider what it is they wish to communicate before they begin writing, or dictating to a secretary. As a result, we rely too much on archaic conventions and courtesies and do not pay sufficient attention to the essential purpose of most letters, which is to request or supply information. Many letters show a reluctance to come to the point. There is frequently a lack of essential or relevant information in letters. There is insincerity and overpoliteness and many letters are much longer than they need be, with unnecessary circumlocutions in construction. These are some of the problems with which we have to deal if we are to read our correspondence efficiently. You might like to list other problems which you feel you encounter in your daily handling of correspondence.

One solution for correspondence lies in **skimming for central points**—with many letters this will tell you where the letter has to go or what *kind* of answer is required or what *kind* of action you have to take. *You need often only read in full those letters* *on which you have to take action yourself.* You may find that, if you take trouble with your own letters to make sure that their purposes are immediately clear to their readers, you will in return receive letters that have been drafted with a little more thought for the problems of readers. At the very least, you will be helping the people with whom you correspond to make a more efficient use of their reading skills. Examine your own handling of correspondence. Would the solution offered here help you to reduce the time taken to deal with it?

JOURNALS AND MAGAZINES

It is here that business and industrial readers probably experience their greatest problems. Many will be on a circulation list for a number of journals relevant to their work and will be expected to use the opportunities which this circulation of material offers to keep abreast of developments in their own fields. Some will not avail themselves of this opportunity because they feel that the size of the reading task is larger than it actually is.

Many readers complain that journals are often bulky and appear too frequently. Many journals appear monthly and if you are on the circulation list for several journals this rate of publication can be too great by far. A further problem is the range of subjects with which you often have to be familiar in order to derive the maximum benefit from reading a journal and even, sometimes, simply to understand much of the material. Frequently the vocabulary is very wide or the material covers several specialized vocabularies. As with newspapers, there is often a bias in the articles, though not necessarily a political bias, and the reader needs to be aware of the writer's preference for a particular viewpoint, so that he can allow for it in his evaluation of the material. Not all relevant information on a subject may be given because a certain amount of prior knowledge may, often unjustifiably, be assumed. You may, again, want to list other problems which you encounter in dealing with journals and magazines.

An effective way of dealing with journals is to **preview the**

whole journal and select the articles to be read. Of each
article selected, **read the synopsis.** If no synopsis is provided,
read the opening and closing few paragraphs and look for
main ideas and important facts in the rest of the article. Read
an article in full only when necessary or desired. It is also
useful to **note the advertising layout in the journal,** so that
you may avoid these pages if you wish. This is easiest to do
where advertisements are placed at the beginning and end of
the journal, with the editorial matter sandwiched between.
Even where advertisements are distributed more or less evenly
through the journal, you can avoid wasting time reading advertisements by learning their whereabouts through previewing.
In some specialized journals, of course, one of your primary
purposes in reading may be to read the advertisements since
they may contain information about new books, products,
and so on.

REPORTS: TECHNICAL, BUSINESS, AND INDUSTRIAL

Here again, there is often a problem of bulk, because many reports
of necessity contain much background material which not every
reader of the report will need to read. Often reports are neither
well constructed nor well written. There is frequently a wide
circulation for reports within an organization and, because
different departments need different information from reports,
this can make for a confused style which tries to appeal to everybody, but in fact helps no one.

Most well-written reports will have a summary, placed at
the beginning, in which the main points and the most important
information will be brought together so that readers can see
quickly how relevant the report is to them personally. Reports
that contain more than about half a dozen pages may have a table
of contents to guide the reader. Longer reports may have an index,
especially if they cover a wide field relevant to a number of
departments in an organization or if information relevant to one

department is featured in several sections of the report. Only if reports are effectively written, according to basic principles such as those outlined here, can readers deal with them efficiently. Badly written reports will lead to inefficient reading unless certain steps are taken.

You will need to preview badly written reports carefully, trying to detect such pattern in the organization as there is. You may have to be particularly ruthless in your approach and read only those parts of the material which clearly concern you. Whenever possible, refer badly written reports back to the writer and ask *him* to summarize the main points of the report as they are likely to affect you. It is also useful if you can persuade your company or organization to adopt a clear policy for the layout and presentation for the main types of reports in circulation. It is easier to write well within a set pattern than to have to create your own pattern on each occasion. Many reports can be written within the following pattern:

1. Summary of the whole report.
2. Table of contents.
3. Introduction.
4. Body of the report.
5. Conclusions.
6. Appendices.

In dealing with reports organized like this, **read the summary first**. This will give you an overview of the whole report. **Read the table of contents** and then **skim through the report as a whole, noting or marking the parts you will return to for a more detailed consideration.** In this skimming, concentrate on looking for main ideas and on becoming aware of the logical development of the material. Then re-read the selected points if necessary, especially if you are reading the report in preparation for a discussion on the report's findings. Approaching reports in this way will enable you both to reduce the time you spend in reading and studying reports and to make your reading of them more productive. Two of the exercises at the end of this chapter are reports, and one of them has more to say about the question of report writing.

LEISURE READING

In this book, we are primarily concerned with reading that you *have* to do rather than with reading that you *want* to do. However, as we have already said, one of the best ways of achieving long-term improvements in the quality of reading comprehension is to read widely. With this in mind, it might be useful for us to look at some of the problems encountered in leisure reading and to offer some suggestions that may enable you to derive more enjoyment and benefit from this reading.

One difficulty in leisure-time reading is the selection of suitable material. Many people have difficulty in deciding whether a book is what they are looking for until they are well on with reading it. This is obviously wasteful of effort and none of us has so much leisure time that we can ignore altogether considerations of time and efficiency. Then there is the problem of finding the time for reading, say, a novel. Many people find that if they are interested in what they are reading, they read more slowly—consequently, it may take them so long to read a novel that they are reluctant to even begin one. Many readers also find it difficult to evaluate what they have read and feel that they have to be content simply with saying that they liked or did not like a particular novel and leave it at that.

Reading Fiction

If you are choosing a work of fiction to read, whether it be a novel, a play, or poetry, it can be a help if you **find out something about the writer first**. In this way, you will be more likely to be able to select material written by people whose experience of and outlook on life has something to contribute to your own. It also helps if you can **give reading a higher priority in your leisure activities** and if you try to read a fairly regular amount each week. In your leisure reading, you should **try to read**

easy and more difficult material in more or less equal amounts.
If you wish to evaluate what you read, in order to give you greater
satisfaction than simply liking or disliking can provide, you should
do this methodically. You will recall that the section in **Chapter 3**
that dealt with Critical Reading offered just such a method, which
can be applied equally usefully to reading at work and to leisure
reading. You will generally gain more from your leisure reading
if you **try to plan ahead a little.** Choose a number of books to be
read each month and try to keep up with this reading program.
Concentrate on reading not only for pleasure but also for a deeper
sense of satisfaction and fulfilment.

Reading Nonfiction

In reading nonfiction, such as biographies and hobby
handbooks, you can again try to be methodical. To select the
best book on a subject, you can **preview several,** using the **table
of contents,** the **index,** and any **chapter summaries** that are
provided. **Read the opening and closing few paragraphs of each
chapter** to get an overview of each book. Incidentally, if you wish
to become more accurate in selecting suitable nonfiction reading
matter, you will find it useful to **learn the system according to
which your local library organizes its books.** Most libraries use
the same system for the classification of books on the shelves
(the Dewey system) and by knowing the code number of each
of the principal subject groupings, you can quickly find those
parts of any library where books that are likely to appeal to you
are kept.

VOCABULARY

You should always have a **dictionary** handy when you are
reading. As you read a book, for example, mark words you do
not understand. Try to guess the meaning, if you can, from the
context in which the word is used. Only consult a dictionary at
this point if you feel you cannot continue to read without some

knowledge of the meaning of the word. When you have finished reading and if the meaning of the word has not become clear, consult a dictionary. *Keep a record of new words learned in this way in a notebook, together with their dictionary definitions and one or two example sentences in which you have used the words yourself.*

RELAXATION

If you are to deal efficiently with your daily reading, it is desirable, especially if you are "working against the clock" and have other important work to do, to avoid tension as you read. By this is meant an undue preoccupation with the fact of having too much to do in too little time. It is possible to become so concerned with the problem of coping with all your reading in the time available that both the speed and quality of your comprehension are seriously impaired. In extreme cases, this feeling of tension manifests itself in a tightening of the muscles at the back of the neck, an increased pulse rate, and even stomach upsets. These symptoms contribute nothing to efficiency in reading and can lead to frequent headaches and to other "stress" conditions.

A little tension, a sense of feeling "tuned up," can have beneficial results occasionally when a particularly important piece of reading is being tackled, but for all normal daily reading it is important to have a relaxed, confident approach. We are not speaking here of relaxation in the more usual physical sense of finding a comfortable chair and putting your feet up. We are, rather, concerned with mental attitudes. There are many ways of achieving a relaxed approach when reading, but probably one of the best is to become so interested in what you are reading that tensions disappear. This should not be too difficult to achieve when you are dealing with reading at work.

ANTICIPATION

Anticipation, too, is a question of mental attitudes. In this connection, it is important to realize that every reader anticipates to a certain extent the nature of the material he has yet to read.

A useful analogy can be made with the skill of driving a car. Even the poorest driver is aware of some of the hazards he is approaching and prepares to take action that is appropriate to the situation. We all ease off the accelerator when we notice that the traffic lights up the road have turned red or when we see children approaching the crossing two junctions ahead, but the good driver is aware of most of the situations he will encounter, even those too far ahead for the poorest driver to have noticed. Thus, he can take decisive, appropriate action in good time to overcome each little obstacle successfully. *In reading we must try to develop our anticipation skills and try to foresee the end toward which the writer is progressing.* Of course, we may often be mistaken at first, but this isn't important—at least we are becoming more actively involved in our reading and this is what matters. *It is active involvement that will encourage the development of accurate anticipation.* So try to be one step ahead of the writer, try to grasp as quickly as possible the overall organization of the material and use this to encourage rapid and efficient reading. A useful aid in the development of anticipation skills is to take selected newspaper articles, for example, and read the opening paragraphs only. Write down what you think the writer is going to tell you (in essence), then read the whole article and check how accurate your anticipation and expectations were.

CONCENTRATION

Your **concentration** as you read should already be better than it was at the beginning of this book because of your **refusal to regress** during practice sessions and also through your **definition of purpose** before you begin reading. Improving your ability to relax as you read and improving your **anticipation skills** will further improve your concentration as you read. In fact, the **active involvement** referred to above is your best aid to improved concentration and, as a result of this, you will also have a **better retention of information** and a **better understanding of what you read.** You should remember to allow better concentration to arise naturally out of all these things and should avoid the intensive, frowning concentration which achieves nothing more than a pain between the eyes.

EXERCISES—NEWSPAPER ARTICLE

Instructions

Preview before reading.

Define your purposes in reading.

Read through the passage once only *as quickly as you can without loss of comprehension.* Do not regress.

Try to improve upon your previous best performance.

Begin timing and begin reading NOW.

INTELLIGENCE: A CHANGED VIEW

by Douglas Pidgeon

The use of intelligence tests has been dominated by the idea that intelligence was an inherited characteristic of the mind that could be reasonably accurately measured. The tests have been used to find out whether children were "working up to capacity." If the Intelligence Quotient was well above the attainment level, then clearly the fault lay with factors other than intelligence and, where possible, steps could be taken to remedy the situation. But if both Intelligence Quotient and attainment were low, then nothing could be done since it was assumed that the children were innately dull.

Intelligence was believed to be a fixed entity, some faculty of the mind that we all possess and which determines in some way the extent of our achievements. Since the Intelligence Quotient was relatively unaffected by bad teaching or a dull home environment, it remained constant, at least within the limits of measurement error. Its value, therefore, was as a predictor of children's future learning. If they differed markedly in their ability to learn complex tasks, then it was clearly necessary to educate them differently—and the need for different types of school and even different ability groups within schools was obvious. Intelligence tests could be used for streaming children according to ability at an early age; and at 11 these tests were superior to measures of attainment for selecting children for different types of secondary education.

Today, we are beginning to think differently. In the last few years, research has thrown doubt on the view that innate intelligence can ever be measured and on the very nature of intelligence itself. Perhaps most important, there is considerable evidence now which shows the great influence of environment both on achievement and intelligence. Children with poor home backgrounds not only do less well in their school work and in intelligence tests—a fact which could be explained on genetic grounds—but their performance tends to deteriorate gradually compared with that of their more fortunate classmates. Evidence like this lends support to the view, stressed by Sir Cyril Burt, that we have to distinguish between genetic intelligence and observed intelligence. Any deficiency in the appropriate genes will obviously restrict development, no matter how stimulating the environment. But we cannot observe or measure innate intelligence; whereas we can observe and measure the effects of the interaction of whatever is inherited with whatever stimulation has been received from the environment. Changes may occur in our observations or measurements if the environment is changed. In other words, the Intelligence Quotient is not constant.

Researches over the past five or ten years, especially in the United States, have been investigating what happens in this interaction. Work in this country has shown that parental interest and encouragement are more important than the material circumstances of the home.

Two major findings have emerged from these studies. Firstly, that the greater part of the development of observed intelligence occurs in the earliest years of life. Professor Bloom in the University of Chicago has estimated that 50 per cent of measurable intelligence at age 17 is already predictable by the age of four. In other words, deprivation

in the first four or five years of life can have greater consequences than any of the following twelve or so years. And
the longer the early deprivation continues the more difficult
it is to remedy.

Secondly, the most important factors in the environment
are language and psychological aspects of the parent-child
relationship. Much of the difference in measured intelligence
between "privileged" and "disadvantaged" children may be
due to the latter's lack of appropriate verbal stimulation and
the poverty of their perceptual experiences.

These research findings have led to a revision in our
understanding of the nature of intelligence. Instead of it
being some largely inherited fixed power of the mind, we
now see it as a set of developed skills with which a person
copes with any environment. These skills have to be learned
and, indeed, one of them—a fundamental one—is learning
how to learn. From birth a baby learns from his environment,
and how to react with it. He learns from one experience how
to cope with other similar experiences and then with
different ones.

It seems equally certain that any built-in mechanism
for learning needs to be sustained and encouraged. As Piaget
has said, a child must play an active part in regulating his own
development; he must be allowed to do his own learning
because full intellectual development will not occur if his role
is a passive one. The more new things he has seen and heard,
the more interested he is in seeing and hearing. The more
different things he has coped with himself, the greater his
capacity for coping. In this way the intellectual skills of
intelligence are built up.

But some children are born into a world where there are
few, if any, of the basic requirements for normal development. Jensen claims that while middle-class children with
low Intelligence Quotient are indeed slow learners, disadvantaged children with low IQ show a wide range of
learning ability. In other words, for them, low IQ is indeed
a poor index of their ability to learn. It might be safest to
assume that the IQ is an underestimate for all children,
although clearly, the more a home is known to have provided appropriate cultural and educational stimulation,

the more likely it is that the IQ does reflect innate potential.

Bloom suggests that most children will master any task or solve any problem provided they are given sufficient time. He admits that a few children, probably less than 5 per cent, may need an impossibly long time to learn some tasks. This fits in well with the changed view of intelligence. High intelligence is not the ability to learn complex tasks so much as the ability to learn rapidly. And a child of relatively low intelligence is not incapable of learning complex tasks but needs a longer time to learn them.

Although educationists have been aware of this idea—the term "slow learner" is sometimes applied to a child of low IQ—the "fixed potential" concept of intelligence has tended to dominate our education system. Teachers are being constantly assured through their classroom experiences that this concept is apparently sound, and that a system which separates children into groups according to their innate potential, really works. This attitude towards the fixed potential concept of intelligence is likely to remain deeply ingrained, since the more successful teachers are at matching pupils' attainments to their apparent abilities, the more successful will any initial streaming or selection process appear to be. This is because the teacher who believes that observed intelligence largely determines the likely level of achievement will strive to see that this happens. Children themselves are obliging creatures and are very inclined to produce the standard of work that their elders regard as appropriate. Thus, streaming and selection procedures are, to some extent, self-fulfilling.

The modern ideas concerning the nature of intelligence put forward here are bound to have some effect on our school system. In one respect a change is already occurring. With the move towards comprehensive education and the development of unstreamed classes, fewer children will perhaps be given the label "low IQ" which must inevitably condemn a child in his own, if not society's eyes. The idea that we can teach children to be intelligent in the same way that we can teach them reading or arithmetic may take some getting used to. But perhaps the greatest changes are still to come.

The greatest gains from this view of intelligence must benefit the disadvantaged child, since there can be no doubt that we have in the past underestimated his potential. Not only must we train him in the skills of learning but if necessary we must make our education system more flexible to give him more time for learning if he needs it. Though all children may be to some extent disadvantaged, some are more so than others.

1300 words

Write down the time taken to read this passage and then attempt the Comprehension Test.

COMPREHENSION TEST

Select the most suitable answer in each case
*Do **not** refer back to the passage.*

A. Retention

1. It used to be believed that intelligence was:
    a) only found in middle-class children.
    b) a variable quantity.
    c) a fixed entity.
    d) only found in children.

2. The greater part of the development of observed intelligence occurs.
    a) in middle-class children.
    b) in the earliest years of life.
    c) in adolescence.
    d) after a person reaches the age of 17.

3. Intelligence is now believed to be:
    a) a set of developed skills.
    b) a fixed entity.
    c) incapable of any kind of measurement.
    d) only found in children.

4. High intelligence is not the ability to learn complex tasks so much as:
   a) the ability to learn rapidly.
   b) the ability to learn a great deal.
   c) the ability to do well in intelligence tests.
   d) the ability to perform tasks older children find difficult.

5. If necessary we must give the slow learner:
   a) a head start on other children.
   b) a better home environment.
   c) more practice in answering intelligence tests.
   d) more time for learning if he needs it.

B. Interpretation

6. The old view of intelligence would seem to justify:
   a) comprehensive education and unstreamed classes.
   b) the selection of children for different types of schools.
   c) sending the duller children out to work at an early age.
   d) taking "disadvantaged" children away from their home environment.

7. A "disadvantaged" child is one:
   a) who comes from a deprived home environment.
   b) who is unintelligent.
   c) who is incapable of learning.
   d) whose parents are poor.

8. A low IQ (Intelligence Quotient) is a poor index of disadvantaged children's ability to learn because:
   a) they may never have really been given a chance to learn.
   b) their intelligence cannot be measured.
   c) they do not come from middle-class homes.
   d) their parents are probably not intelligent.

9. A "slow learner" is:
   a) a polite term for someone with a low IQ.
   b) an unintelligent child who tries hard but does not quite succeed.

     c) a child who may be as "intelligent" as others but takes longer to learn.
     d) a child who just does not do well in intelligence tests.

10.   Teachers are likely to:
     a) welcome the new ideas about the nature of intelligence.
     b) reject completely the new ideas about intelligence.
     c) remain skeptical about the new view of intelligence.
     d) take some time before they generally accept the new view of intelligence.

Convert the time taken into "words per minute" by means of the Reading Speed Conversion Table on page 254. *Enter the result on the Progress Graph on page* 256.

Check your answers to the Comprehension Test against the answers given on page 260. *Enter the result on the Progress Graph on page* 256.

C. **Discussion**—discuss one of these questions (orally if in a group, in writing if studying alone). *You may refer back to the passage.*

11.   What is intelligence?

12.   Can intelligence be measured?

13.   Are animals intelligent?

14.   What are likely to be the effects of these new views of intelligence upon methods of educating children?

15.   Why are the first five years of life so important in deciding the extent of individual development?

Proceed to the next reading exercise.

MAGAZINE ARTICLE

Instructions

Preview before reading.
Define your purposes in reading.

Read through the passage once only *as quickly as you can without loss of comprehension.* Do not regress.

Try to improve upon your previous best performance.

Begin timing and begin reading NOW.

THE THINK-TANK PREDICTORS

by Norman Moss

A new profession has grown up in the United States in the past two decades: thinking. Like any other profession, it has its own methods, its own language and its institutions. These are the private research corporations, or "think tanks" (in the current American terminology), which are paid, mostly, by departments of Government, to think about problems.

There, people think purposefully and, they would say, systematically. They have developed methods of thinking that are tools of their profession—"tool" is a favourite term. Among some of them, one finds the sense that they are "new men," who are learning to use the human brain rationally for the first time in history. These tend to exaggerate the exclusiveness of their craft, and to look upon outsiders who have ideas as mere brooders, amateur thinkers.

Their methods have been applied with the most drastic results to defence problems. The most famous and influential of the think tanks, the Rand Corporation of California, gave the American Government its ideas about the use and non-use of nuclear weapons. However, this is only because Government money has been most readily available in the defence area. The methods can be applied to business or any other situations, and they will be applied to the racial and social problems that are tearing apart American cities.

The most famous alumnus of Rand is Herman Kahn, author of three books on nuclear war and peace, inventor of the term "escalation" and widely (and wrongly) believed to be the model for Dr. Strangelove. He left Rand in 1961 to found his own think tank, the Hudson Institute in the wooded countryside a few miles up the Hudson River from New York City.

As well as a talented and imaginative instrumentalist on a computer, Kahn is a humorous, ebullient, corpulent extrovert, who gushes ideas on an extraordinary variety of subjects. As an author and lecturer, he has done much to popularise some of the salient ideas of the think tanks, and colleagues' respect for his intellect is often tinged with that slight reserve that scholars feel for one who popularises their scholarship.

Recently, Kahn has turned to exploring the future—the remaining years of this century. With the help of a grant from a philanthropic foundation he has turned on them the practised minds and rigorous methods of the Hudson Institute. Now he and another member of the Hudson Institute, Anthony Wiener, have published the cautious conclusions of these studies. They reflect at the same time Kahn's gargantuan intellectual appetite, and the characteristic think-tank attitude that everything can be made the subject of calculations. Using charts, graphs and extrapolations from current tendencies, they speculate on the future of contraception as well as economic growth, and alienation as well as international relations.

Kahn and Wiener have constructed what they call "surprise-free projections," which are arrived at by extrapolating current trends. Out of these, they construct a picture of what they call the year 2000, which seems to them more likely than any other single one. Then they give plausible variations on this. Theirs is the calm, non-apocalyptic school of predictions. In the surprise-free world, the international scene turns out not to be very different from the present one. Kahn and Wiener acknowledge this: "We would be willing to wager small sums, at even odds that the next third of the century will contain fewer big surprises than either of the previous thirds (i.e., that in this respect, the world is more like 1815 than 1914)."

In the next thirty-two years, it seems the rich nations will get richer and the poor ones will get richer too, but, still being behind, will still be discontented. Britain, creaking at the same social and economic joints that slow her growth today, will grow richer at a slower pace than some other countries, including her nearest Continental neighbours, and so will fall behind them. Only Japan will change her position, becoming relatively richer and more powerful so that she will rank just below the two super-powers.

Kahn and Wiener list 100 technological innovations that are "very likely" by that time, and 25 more that are "less likely but important possibilities." On the first list comes, or is coming, the ability to choose the sex of an unborn child, a computer to run a household, drugs that can ensure that a person sleeps, and dreams, when and how he wants, others to regulate effortlessly his—or her—weight, and artificial moons to light up whole areas of the earth at night. Some items on the second list are sobering to contemplate a mere half-a-lifetime away: suspended animation for years or centuries; extension of human life span to between 100 and 150 years; an increase in mental capacity to be achieved by connecting a computer directly to the brain; modification of the solar system.

But (in common with good science fiction writers) Kahn and Wiener are more concerned with human character and values than with gadgetry, and they point to changes in the role that work plays in people's lives and ideas of status when it is no longer so necessary to earn a living. Without this necessity to nail people down to objective reality, they say, fads, exotic fashions and cults will have large followings. Very many youngsters will just "bum around" which will become easier when there is still more money about, and some of them will "cloak themselves in the pretensions of artistic creativity." To this extent, King's Road and Ibiza are the wave of the future.

This, then, is the surprise-free world; what happens when history injects its usual surprises? Kahn and Wiener offer one non-standard world in which Communish is on the retreat everywhere, another with a United Europe including most of what is now the Eastern Bloc, a third with a world-wide economic depression. In another, Chinese support for a revolution in Mexico leads to a limited US-Chinese nuclear exchange. In yet another an elitist, technocratic, pan-European movement starts in Spain of all places and gives Europe a new dynamism and militancy.

The authors worry a lot about totalitarianism backed by new technology; some of their darker pictures of the year

2000 resemble that other one of the year 1984. "Bugging" could be automated, so that a monitoring device would react to certain key phrases like "black power," "infiltrate" or "organise," or simply to a rising tone of emotion in the voice. People's states of mind might be controlled by drugs.

Despite the rigour of their methods, there are limits to objectivity in prediction. Ultimately, acceptance of the Kahn-Wiener view of the future must rest on their picture of the present, and this, though consensual, is not universal. However, their projections are both stimulating and plausible. They may make some of the surprises a little less surprising. And, however straight the surprise-free track, the world's stockpile of nuclear bombs lies along some of the branch lines. We are still under that ancient Chinese curse, "May you live in interesting times."

1200 words

Write down the time taken to read this passage and then attempt the Comprehension Test.

COMPREHENSION TEST

Select the most suitable answer in each case.
Do not refer back to the passage.

A. Retention

1. A "think tank" is another name for:
   a) a private research corporation.
   b) a private swimming pool.
   c) an imaginary weapon of war.
   d) a computer.

2. Kahn and Wiener believe that the last third of the century will contain:
   a) many big surprises.
   b) more big surprises.

    c) fewer big surprises.

    d) no big surprises.

3. Kahn and Wiener believe that the only country that will improve its economic position relative to others is:

    a) USA.

    b) China.

    c) Britain.

    d) Japan.

4. In one of the "worlds" of the year 2000 forecast by Kahn and Wiener, there is a limited nuclear exchange between:

    a) USA and USSR.

    b) USA and China.

    c) China and USSR.

    d) USA and Britain.

5. The writer says we are still under the ancient Chinese curse:

    a) "May you live in interesting times."

    b) "May you live to be a hundred."

    c) "May you never be surprised."

    d) "May you live in unsettled times."

B. Interpretation

6. An alumnus is:

    a) a rare metal.

    b) a writer.

    c) a former student

    d) a scaremonger.

7. Ebullient means:

    a) exuberant.

    b) fat.

    c) bullying.

    d) pessimistic.

8. A "surprise-free" projection is:

    a) a forecast of what is certain to happen.

    b) an unpleasant picture of the future.

    c) one based only on what has happened in the past.

    d) an attempt to offer a realistic and accurate picture of the future.

9. The kind of world that Kahn and Wiener predict for 2000
   can best be described as:
   a) idyllic.
   b) a return to the past.
   c) much the same as today's world.
   d) of unimaginable horror.

10. The greatest advances in the next thirty years can,
    according to Kahn and Wiener, be expected in:
    a) science and technology.
    b) politics and economics.
    c) leisure and the arts.
    d) social life.

*Convert the time taken into "words per minute" by means
of the Reading Speed Conversion Table on page 254. Enter the
result on the Progress Graph on page 256.*

*Check your answers to the Comprehension Test against the
answers given on page 260. Enter the result on the Progress Graph
on page 256.*

C. **Discussion**—discuss one of these questions (orally if in a group,
   in writing if studying alone). *You may refer back
   to the passage.*

11. Which of the futures offered by Kahn and Wiener do you
    think most likely?

12. Can there be any such thing as a "surprise-free pro-
    jection" of the future?

13. Does the work of the "think tanks" serve any useful
    purpose?

14. How do Kahn's and Wiener's views of the future compare
    with those of either
    a) Aldous Huxley in *Brave New World*? or
    b) George Orwell in *1984*? or
    c) the views presented in any other novel, play, film, or
       other work of fiction of your own choice?

15. What are your own views of what the world will be like
    in the year 2000?

Proceed to the next reading exercise.

JOURNAL ARTICLE

Instructions

Preview before reading.

Define your purposes in reading.

Read through the passage once only *as quickly as you can without loss of comprehension.* **Do not regress.**

Try to improve upon your previous best performance.

Begin timing and begin reading NOW.

COMMUNICATION WITHOUT WORDS*

by *Albert Mehrabian*

Suppose you are sitting in my office listening to me describe some research I have done on communication. I tell you that feelings are communicated less by the words a person uses than by certain nonverbal means—that, for example, the verbal part of a spoken message has considerably less effect on whether a listener feels liked or disliked than a speaker's facial expression or tone of voice.

So far so good. But suppose I add, "In fact, we've worked out a formula that shows exactly how much each of these components contributes to the effect of the message as a whole. It goes like this: Total Impact = .07 verbal + .38 vocal + .55 facial."

What would you say to *that*? Perhaps you would smile good-naturedly and say, with some feeling, "Baloney!" Or perhaps you would frown and remark acidly, "Isn't science grand." My own response to the first answer would probably be to smile back: the facial part of your message, at least, was positive (55 percent of the total). The second answer

*Reprinted by permission of *Psychology Today Magazine.* Copyright ©1968 Ziff-Davis Publishing Company.

might make me uncomfortable: only the verbal part was positive (7 percent).

The point here is not only that my reactions would lend credence to the formula but that most listeners would have mixed feelings about my statement. People like to see science march on, but they tend to resent its intrusion into an "art" like the communication of feelings, just as they find analytical and quantitative approaches to the study of personality cold, mechanistic and unaccaptable.

The psychologist himself is sometimes plagued by the feeling that he is trying to put a rainbow into a bottle. Fascinated by a complicated and emotionally rich human situation, he begins to study it, only to find in the course of his research that he has destroyed part of the mystique that originally intrigued and involved him. But despite a certain nostalgia for earlier, more intuitive approaches, one must acknowledge that concrete experimental data have added a great deal to our understanding of how feelings are communicated. In fact, as I hope to show, analytical and intuitive findings do not so much conflict as complement each other.

It is indeed difficult to know what another person really feels. He says one thing and does another, he seems to mean something but we have an uneasy feeling it isn't true. The early psychoanalysts, facing this problem of inconsistencies and ambiguities in a person's communications, attempted to resolve it through the concepts of the conscious and the unconscious. They assumed that contradictory messages meant a conflict between superficial, deceitful, or erroneous feelings on the one hand and true attitudes and feelings on the other. Their role, then, was to help the client separate the wheat from the chaff.

The question was, how could this be done? Some analysts insisted that inferring the client's unconscious wishes was a completely intuitive process. Others thought that some nonverbal behavior, such as posture, position and movement, could be used in a more objective way to discover the client's feelings. A favorite technique of Frieda Fromm-Reichmann, for example, was to imitate a client's posture herself in order to obtain some feeling for what he was experiencing.

Thus began the gradual shift away from the idea that communication is primarily verbal, and that the verbal message includes distortions or ambiguities due to unobservable motives that only experts can discover.

Language, though, can be used to communicate almost anything. By comparison, nonverbal behavior is very limited in range. Usually, it is used to communicate feelings, likings and preferences, and it customarily reinforces or contradicts the feelings that are communicated verbally. Less often, it adds a new dimension of sorts to a verbal message as when a salesman describes his product to a client and simultaneously conveys, nonverbally, the impression that he likes the client.

A great many forms of nonverbal behavior can communicate feelings: touching, facial expression, tone of voice, spatial distance from the addressee, relaxation of posture, rate of speech, number of errors in speech. Some of these are generally recognized as informative. Untrained adults and children easily infer that they are liked or disliked from certain facial expressions, from whether (and how) someone touches them, and from a speaker's tone of voice. Other behavior, such as posture, has a more subtle effect. A listener may sense how someone feels about him from the way the person sits while talking to him, but he may have trouble identifying precisely what his impression comes from.

Correct intuitive judgments of the feelings or attitudes of others are especially difficult when different degrees of feeling, or contradictory kinds of feeling, are expressed simultaneously through different forms of behavior. As I have pointed out, there is a distinction between verbal and vocal information (vocal information being what is lost when speech is written down—intonation, tone, stress, length and frequency of pauses, and so on), and the two kinds of information do not always communicate the same feeling. The distinction, which has been recognized for some time, has shed new light on certain types of communication. Sarcasm, for example, can be defined as a message in which the information transmitted vocally contradicts the information transmitted verbally. Usually the verbal information is positive and the vocal is negative, as in "Isn't science grand."

Through the use of an electronic filter, it is possible to measure the degree of liking communicated vocally. What the filter does is eliminate the higher frequencies of recorded speech, so that words are unintelligible but most vocal qualities remain. (For women's speech, we eliminate frequencies higher than about 200 cycles per second; for men, frequencies over about 100 cycles per second.) When people are asked to judge the degree of liking conveyed by the filtered speech, they perform the task rather easily and with a significant amount of agreement.

This method allows us to find out, in a given message, just how inconsistent the information communicated in words and the information communicated vocally really are. We ask one group to judge the amount of liking conveyed by a transcription of what was said, the verbal part of the message. A second group judges the vocal component, and a third group judges the impact of the complete recorded message. In one study of this sort we found that, when the verbal and vocal components of a message agree (both positive or both negative), the message as a whole is judged a little more positive or a little more negative than either component by itself. But when vocal information contradicts verbal, vocal wins out. If someone calls you "honey" in a nasty tone of voice, you are likely to feel disliked; it is also possible to say "I hate you" in a way that conveys exactly the opposite feeling.

Besides the verbal and vocal characteristics of speech, there are other, more subtle, signals of meaning in a spoken message. For example, everyone makes mistakes when he talks—unnecessary repetitions, stutterings, the omission of parts of words, incomplete sentences, "ums" and "ahs." In a number of studies of speech errors, George Mahl of Yale University has found that errors become more frequent as the speaker's discomfort or anxiety increases. It might be interesting to apply this index in an attempt to detect deceit (though on some occasions it might be risky: confidence men are notoriously smooth talkers).

Timing is also highly informative. How long does a

speaker allow silent periods to last, and how long does he wait before he answers his partner? How long do his utterances tend to be? How often does he interrupt his partner, or wait an inappropriately long time before speaking? Joseph Matarazzo and his colleagues at the University of Oregon have found that each of these speech habits is stable from person to person, and each tells something about the speaker's personality and about his feelings toward and status in relation to his partner.

Utterance duration, for example, is a very stable quality in a person's speech; about 30 seconds long on the average. But when someone talks to a partner whose status is higher than his own, the more the high-status person nods his head the longer the speaker's utterances become. If the high-status person changes his own customary speech pattern toward longer or shorter utterances, the lower-status person will change his own speech in the same direction. If the high-status person often interrupts the speaker, or creates long silences, the speaker is likely to become quite uncomfortable. These are things that can be observed outside the laboratory as well as under experimental conditions. If you have an employee who makes you uneasy and seems not to respect you, watch him the next time you talk to him—perhaps he is failing to follow the customary low-status pattern.

Immediacy or directness is another good source of information about feelings. We use more distant forms of communication when the act of communicating is undesirable or uncomfortable. For example, some people would rather transmit discontent with an employee's work through a third party than do it themselves, and some find it easier to communicate negative feelings in writing than by telephone or face to face.

Distance can show a negative attitude toward the message itself, as well as toward the act of delivering it. Certain forms of speech are more distant than others, and they show fewer positive feelings for the subject referred to. A speaker might say, "Those people need help," which is more distant than "These people need help," which is in turn even more distant than "These people need our help." Or he might say "Sam and I have been having dinner," which has less immediacy than "Sam and I are having dinner."

Facial expression, touching, gestures, self-manipulation (such as scratching), changes in body position, and head movements—all these express a person's positive and negative attitudes, both at the moment and in general, and many reflect status relationships as well. Movements of the limbs and head, for example, not only indicate one's attitude toward a specific set of circumstances but relate to how dominant, and how anxious, one generally tends to be in social situations. Gross changes in body position, such as shifting in the chair, may show negative feelings toward the person one is talking to. They may also be cues: "It's your turn to talk," or "I'm about to get out of here, so finish what you're saying."

Posture is used to indicate both liking and status. The more a person leans toward his addressee, the more positively he feels about him. Relaxation of posture is a good indicator of both attitude and status, and one that we have been able to measure quite precisely. Three categories have been established for relaxation in a seated position: least relaxation is indicated by muscular tension in the hands and rigidity of posture; moderate relaxation is indicated by a forward lean of about 20 degrees and a sideways lean of less than 10 degrees, a curved back, and, for women, an open arm position; and extreme relaxation is indicated by a reclining angle greater than 20 degrees and a sideways lean greater than 10 degrees.

Our findings suggest that a speaker relaxes either very little or a great deal when he dislikes the person he is talking to, and to a moderate degree when he likes his companion. It seems that extreme tension occurs with threatening addressees, and extreme relaxation with nonthreatening, disliked addressees. In particular, men tend to become tense when talking to other men whom they dislike; on the other hand, women talking to men *or* women and men talking to women show dislike through extreme relaxation. As for status, people relax most with a low-status addressee, second-most with a peer, and least with someone of higher status than their own. Body orientation also shows status:

in both sexes, it is least direct toward women with low status and most direct toward disliked men of high status. In part, body orientation seems to be determined by whether one regards one's partner as threatening.

The more you like a person, the more time you are likely to spend looking into his eyes as you talk to him. Standing close to your partner and facing him directly (which makes eye contact easier) also indicate positive feelings. And you are likely to stand or sit closer to your peers than you do to addressees whose status is either lower or higher than yours.

What I have said so far has been based on research studies performed, for the most part, with college students from the middle and upper-middle classes. One interesting question about communication, however, concerns young children from lower socioeconomic levels. Are these children, as some have suggested, more responsive to implicit channels of communication than middle- and upper-class children are?

Morton Wiener and his colleagues at Clark University had a group of middle- and lower-class children play learning games in which the reward for learning was praise. The child's responsiveness to the verbal and vocal parts of the praise-reward was measured by how much he learned. Praise came in two forms: the objective words "right" and "correct," and the more affective or evaluative words, "good" and "fine." All four words were spoken sometimes in a positive tone of voice and sometimes neutrally.

Positive intonation proved to have a dramatic effect on the learning rate of the lower-class group. They learned much faster when the vocal part of the message was positive than when it was neutral. Positive intonation affected the middle-class group as well, but not nearly as much.

If children of lower socioeconomic groups are more responsive to facial expression, posture and touch as well as to vocal communication, that fact could have interesting applications to elementary education. For example, teachers could be explicitly trained to be aware of, and to use, the forms of praise (nonverbal or verbal) that would be likely to have the greatest effect on their particular students.

Another application of experimental data on communication is to the interpretation and treatment of schizophrenia. The literature on schizophrenia has for some time emphasized that parents of schizophrenic children give off contradictory signals simultaneously. Perhaps the parent tells the child in words that he loves him, but his posture conveys a negative attitude. According to the "double-bind" theory of schizophrenia, the child who perceives simultaneous contradictory feelings in his parent does not know how to react: should he respond to the positive part of the message, or to the negative? If he is frequently placed in this paralyzing situation, he may learn to respond with contradictory communications of his own. The boy who sends a birthday card to his mother and signs it "Napoleon" says that he likes his mother and yet denies that he is the one who likes her.

In an attempt to determine whether parents of disturbed children really do emit more inconsistent messages about their feelings than other parents do, my colleagues and I have compared what these parents communicate verbally and vocally with what they show through posture. We interviewed parents of moderately and quite severely disturbed children, in the presence of the child, about the child's problem. The interview was video-recorded without the parents' knowledge, so that we could analyze their behavior later on. Our measurements supplied both the amount of inconsistency between the parents' verbal-vocal and postural communications, and the total amount of liking that the parents communicated.

According to the double-bind theory, the parents of the more disturbed children should have behaved more inconsistently than the parents of the less disturbed children. This was not confirmed: there was no significant difference between the two groups. However, the *total amount* of positive feeling communicated by parents of the more disturbed children was less than that communicated by the other group.

This suggests that (1) negative communications toward disturbed children occur because the child is a problem and therefore elicits them, or (2) the negative attitude precedes the child's disturbance. It may also be that both factors operate together, in a vicious circle.

If so, one way to break the cycle is for the therapist to create situations in which the parent can have better feelings toward the child. A more positive attitude from the parent may make the child more responsive to his directives, and the spiral may begin to move up instead of down. In our own work with disturbed children, this kind of procedure has been used to good effect.

If one puts one's mind to it, one can think of a great many other applications for the findings I have described, though not all of them concern serious problems. Politicians, for example, are careful to maintain eye contact with the television camera when they speak, but they are not always careful about how they sit when they debate another candidate of, presumably, equal status.

Public relations men might find a use for some of the subtler signals of feeling. So might Don Juans. And so might ordinary people, who could try watching other people's signals and changing their own, for fun at a party or in a spirit of experimentation at home. I trust that does not strike you as a cold, manipulative suggestion, indicating dislike for the human race. I assure you that, if you had more than a transcription of words to judge from (7 percent of total message), it would not.

2800 words

COMPREHENSION TEST

Summary

In about 250 of your own words, summarize the main points made in this article. Do not *refer back to the passage.*

Note

You may enter your reading speed on the progress graph and may also, if you wish, enter an estimate of your comprehension level on this passage. Proceed to the next reading exercise.

REPORT

Instructions

Preview before reading.

Define your purposes in reading.

Read through the report that follows once only *as quickly as you can without loss of comprehension.* Do not regress.

Try to improve upon your previous best performance.

Begin timing and begin reading NOW.

REPORT WRITING

*A Brief Guide to Effective Report Writing,
Presented in Report Form*

by G. R. Wainwright

1. SUMMARY:

The quality of industrial report writing is frequently inadequate for the demands made upon it, and concern is felt in many quarters about the absence of any real attempt, nationally, to improve it. Particular colleges and firms in particular localities have, however, made some attempts to solve the problem. This is one of those attempts. Whatever success it may have will depend to a large extent on the desire of the report writers themselves to improve their powers and fluency of self-expression.

Certain general principles are described here which seem to be basic to good report writing.
They are:

(a) The writer must know as accurately as possible what kind of report is required and the full purpose of that report, before he begins collecting the material for the report.

(b) Every report should follow a suitable plan and headings should be used to describe and distinguish each of the sections and subsections.

(c) The writer should only include in his report such material as is necessary and relevant to his purpose.

(d) The report should be written within as short a time as possible, the body of the report being written first and the summary being written last.

(e) After writing the report, the writer should revise and criticize his work until he is completely satisfied that it fulfils the purpose for which it was intended, that it says only what he wants it to say and that it contains no ambiguities or inaccuracies.

Six appendices accompany this report which expand on certain of the points made or supplement what is said in some other way. They include a special appendix containing information and advice on the oral presentation of reports and on oral communication in public generally.

TABLE OF CONTENTS:

2. INTRODUCTION:

In view of the importance of having well-written and effectively presented reports in business and industry, with the resultant savings in time spent reading, some instruction in how to write reports is necessary and an attempt is made to provide this here.

This report explains the general principles of effective report writing and the reader is expected to be able to translate the general principles for himself to meet the needs of particular situations.

The report form was chosen for this work so that the general principles discussed could, at the same time, be demonstrated. For this reason, the report should provide a useful source of reference for business and industrial report writers.

3. PURPOSES OF REPORTS:

It is essential that before any of the work necessary for the report is attempted, the writer should have a clear and accurate understanding of the purpose of the report—that is, he must know the kind of report required, the material it should contain, why the information, together with any comments or recommendations asked for, is needed, and for whom the report is being written. All these must be known before the writer can see the report required in its full context, and aim to fit the report into that context.

(a) Kinds of Reports

Reports can be classified either by contents and function, or by form. The nature of these will affect the nature of the report as a whole. For instance, if reference is made to Appendix I, it will be clearly seen that there is a great deal of difference between a preliminary advisory report on a projected piece of work and a final progress report on work completed. The difference will affect almost every aspect of the reports (especially the style of writing) and they will only be alike in the methods of planning and presentation used. Again, referring to Appendix II, the differences between an informal memo report and a long formal technical report are clearly perceptible. They may each follow a similar basic pattern, but there will be little similarity in, for example, the choice of words and the amount of information given.

(b) Contents of Reports

Little can be said in a general way about the amount and type of information a report should contain, since this will differ with each report. It is perhaps worth remembering here that the writer needs to know why the information, together with any comments or recommendations requested, is needed so that he is better equipped to cut out the irrelevant and to place proper emphasis on essential points.

(c) Readers of Reports

Reports usually move in one or more of four directions—that is, they are read either by the writer's subordinates, his colleagues, his superiors, or customers of the firm. Clearly, he should not allow any written communication to pass from himself to any of these readers unless he is absolutely sure that it is not only grammatically correct, but is also concise and relevant to the subject and the purpose for which it was required. If it does not possess these qualities, it is not likely to be effective. The terms in which the report is written, the amount of information it contains, and the degree of complexity of the report as a whole will change considerably according to the type of reader for whom it is intended. Here, it should be noted that a report may travel in more than one of the four directions indicated, and this should be borne in mind when the report is written.

4. PLANNING A REPORT:

Reports are written in four stages:

Stage 1—The period of preliminary study.

Stage 2—The period of careful planning.

Stage 3—The period of rapid composition.

Stage 4—The period of criticism and revision, followed by a rewriting of the report if this is necessary.

In this section we shall show the writer how to deal with the first two stages.

Every report should be written according to a pre-determined plan. It does not really matter what form this plan takes, provided it is one in which the reader is led logically and smoothly from step to step, and one that includes all that is relevant and necessary. Certain general principles of planning can be laid down, however.

(a) Gathering Material

This is the first stage of the procedure to be followed in writing a report. It usually refers to the period of reading and research that precedes a theoretical report, but in most of the cases with which we are concerned here, the material that will go into the report will be the result of practical work. Tables of results, graphs, notes made when the work was done, and any other information about the subject of the report, are all in the hands of the writer. Only when he has all the material he needs can he take the next step. Any incompleteness in any part of the material will be a severe handicap.

(b) Selection of Material

Now follows the second stage in the writing of a report. Assuming a writer has all the material to hand, he must now select the information he needs for his report. Rarely will he find that all the material acquired is needed, and so anything that is not directly relevant to the subject or that is not wanted for other reasons, must be discarded. Information must not be retained for sentimental reasons (e.g., much hard work was given to discovering it), but only if it is necessary to the report.

(c) Arrangement of Material

The writer now has his report ready, but not in any presentable form, for it has to be arranged in some kind of order before it can be submitted to a reader. In Appendix III, there is an outline of an "ideal" arrangement. Few reports will fit neatly into this plan, but most will comprise the basic elements:

    (i) Summary.
    (ii) Introduction.
    (iii) The Body of the Report.
    (iv) Conclusion.

It is for the writer to choose the arrangement that is most suited to his particular report. To repeat, it does not really matter what that arrangement is as long as it is there and as long as it is obvious to the reader. This is quite easily achieved by the use of headings. Each section of the report should be given a

PRIMARY HEADING

and each subsection should be given a

Secondary Heading

Any further divisions or subsections that are necessary can be indicated by setting in the

Subheading

by about ½ inch and by not underlining it. If desired, headings and subheadings can be numbered or lettered, but this should be done in such a way that each is clearly distinguishable. For example:

1. PRIMARY HEADING

(a) Secondary Heading

    (i) Subheading.
    (ii) Subheading.

The use of a simple, clear system such as this will make the report not only easier to read, but also much easier to write since the arrangement of material under each heading will be made as easy as the arrangement of the material as a whole. Everything will tend to fall much more naturally into its logical position in the report.

5. WRITING A REPORT:

Having carefully planned what he is going to do, the writer now comes to the point where he has to realize the plan. It is the third stage in the writing of a report.

(a) Rapid Composition

It is essential that the report should be written in as short a time as possible. This is helpful to the writer since he is more likely to have the same concept of the report in his mind when he finishes writing it than if he takes some considerable time over the writing. If he changes his idea of what the report should be like in the middle of writing it, the finished report is less likely to be a logical and complete whole and is more likely to confuse the reader by the changes in intention or emphasis.

(b) Order of Writing

As a general rule, the body of the report should be written first for similar reasons to those outlined above. Then the introduction, conclusion, and any other parts of the report should be tackled. The synopsis (summary or abstract) should be the last part of the report to be written. Nothing can be adequately summarized until it is before the writer in its entirety.

(c) Keeping to the Plan

The writer should follow his predetermined plan when writing the report, and should not change it without good reason. It is better to start again than to change direction in the middle of the report, since any change is more likely to create confusion than dispose of it.

(d) Effective Writing

As far as the actual manner of expression or style of writing is concerned, the writer should try to follow the principles outlined in Appendix IV, paying particular attention to the fact that an additional aid to good writing is for the writer to try to keep the potential reader in mind throughout the whole course of writing. Any writer who is unsure of his ability to say exactly what he means, or who lacks confidence in the use of the English language, should take steps to improve his fluency and quality of expression by a course in English Language or in Report Writing. Both methods should be accompanied by wide general reading. There is no shortcut to the effective use of language, but older readers will realize more than younger ones that it pays to be able to speak and write effectively.

(e) Length of Report

No rules can be laid down here. The length will depend on how much the writer has to say, and how difficult it is to say it. Generally speaking, a report should be as short as possible without omitting anything that is relevant or necessary.

(f) Illustrative Material

Unless they are either brief, compact, or an integral part of the section concerned, graphs, statistics, results, and other illustrative material should be relegated to an appendix to the report where they can easily be found when required.

6. PREPARING A REPORT FOR CIRCULATION:

This fourth stage in report writing—revising and criticizing, and making all the other preparations for submission and circulation—is by no means the least important. It is a mistake to think that once a report is written that is the end of it. In some ways, it is only the beginning.

(a) Revising and Criticizing the Report

After completing the report, the writer should check and recheck for any irrelevancies. illogicalities, inaccuracies, faults in expression, and any other inadequacies. It may be helpful if he asks himself the kind of questions suggested in Appendix V. This is a task that is at least as important to an effective report as the writing itself. If necessary, a writer should rewrite a report, rather than submit it in a form with which he is not completely satisfied.

(b) Presentation

Some informal reports may quite easily be submitted to the reader in handwritten form, but most industrial reports will be typewritten. The reason for this may not be a reflection on the writer's legibility, but merely that usually several copies are required for circulation. Since the original manuscript has to be given to a typist for this to be done, it is worth the writer's while checking that she knows exactly how the report is to be laid out in typewritten form. It is also advisable to let her have the manuscript early enough for any necessary corrections to be made before the report has to be submitted.

(c) Submission

The prompt submission of a report to the right person does much to facilitate clear and effective industrial communication, the taking of decisions, and the formulation of company policies. The writer should, therefore, have a clear idea of who will require the completed report, and by what date he will require it.

7. CONCLUSIONS:

As far as report writing in an industrial context is concerned, certain points stand out and deserve particular attention:

(a) The nature and purpose of the report must be known by the writer in advance.

(b) A report may be read by more than one type of reader and this should be remembered by the writer.

(c) Every report should be written according to a plan that is logical and suitable to the subject in hand.

(d) In selecting the material that is to go into the report, the writer must take special care only to include such information as is necessary and relevant to his purpose.

(e) The writer should use a simple and clear system of headings for the sections and subsections of his report so that the reader can find any particular item of information readily and without unnecessary searching.

(f) The report should be written in as short a time as possible, to preserve continuity of thought and style. The body of the report should be written first, then the other parts of the report and finally the summary.

(g) After the report is written, it should be revised and checked for irrelevancies, inaccuracies, grammatical faults, and any other deficiencies.

If the advice and information given here is used and is accompanied by a desire to improve ability to express oneself fluently and effectively, then some small progress may have been made toward solving a few of the problems of industrial communication.

8. APPENDICES

I. *Reports: Classified by Contexts and Function*

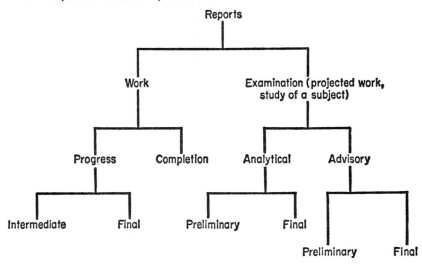

II. *Reports: Classified by Form*

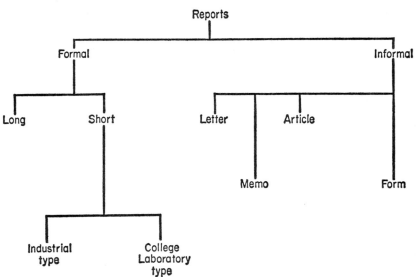

III. *Arrangement of Material for a Report*

(a) Covering letter—polite gesture for formal reports—states the authorization—precedes the title page.

(b) The Title Page—reports of six or more pages.

(c) Summary—a summary of the entire report.

(d) Table of Contents—longer and more formal reports.

(e) Introduction:
   (a) states the subject.
   (b) indicates purpose.
   (c) announces plan of treatment.

(f) Body of the Report—character of investigation, equipment used, procedure followed, detailed results obtained, analysis of results leading to conclusions, recommendations. Sectional headings, paragraphs.

(g) Appendix—relieves body of report of congestion.
   Bibliography—sources of information.

IV. *The Elements of Effective Writing*

(a) Accuracy: get your facts and figures right.

(b) Clarity: sort them out and arrange logically.

(c) Simplicity: within limits imposed by information.

(d) Brevity: within limits imposed by information.

(e) Readability: try to catch and hold the reader's interest.

Seven Rules to Make Writing Effective

(a) Eliminate unnecessary words and keep down the number of long words.

(b) Use direct statements and prefer the active voice to the passive.

(c) Keep ideas per sentence low.

(d) Keep sentences short.

(e) Use qualifying phrases sparingly.

(f) Be specific, rather than general and abstract.

(g) Use adjectives and adverbs with care.

V. *Final Analysis of Written Work*

(a) Are there any unnecessary repetitions?

(b) Is each section of the subject complete in itself?

(c) Is each section in its proper place?

(d) Is any section irrelevant or does any require different planning?

(e) Is the argument well built up?

(f) Have any details been overlooked?

(g) Is there any ambiguity?

(h) Have you kept the reader in mind and tried to persuade *him*?

(i) Is the whole arrangement exactly suited to the purpose?

(j) Does it read smoothly and is it a complete and logical whole?

VI. *Oral Reports*

Basically, there is little difference between a written report and an oral report. What differences there are lie in the delivery of the report, rather than in the stages that precede this. The work on which an oral report is based (the information collected) may well be the same as that for a written report.

An oral report must be planned in a similar way to a written report dividing itself into basic sections:

(a) Introduction.

(b) Body of the Report.

(c) Conclusion.

To put it simply, in the introduction you will tell your audience what you are going to speak about. In the body of the report you will tell them accurately and logically all that you want to say about the particular subject, and in the conclusion you will sum up what you have said, picking out the most important points for further emphasis and repetition.

The main differences between written and oral reports

lie in the expression—that is, in the choice of words and in the manner in which you speak. In a written report, slang and colloquial expressions are completely inappropriate but this may not be the case with all oral reports. The best advice is that you should talk to your audience in the manner that seems most natural and most likely to establish a favorable relationship between them and yourself. Unless this is done you will have the feeling that you have not secured the full attention of your audience, and this is bound to affect your ability to put across effectively what you have to say.

Oral reports or lectures should never be read. You should make notes on what you want to say, using similar headings to those you would use in a written report, and use them as a guide to what you tell your audience. The real purpose of the notes lies in preventing you from forgetting any important part of your report, rather than in providing you with the actual words to use. No attempt should be made to memorize the report (this will make it sound unnatural, dull, and uninteresting), though it does help if you can give a "practice" talk beforehand when alone, or in the presence of a friend who can be trusted to listen and possibly criticize and suggest improvements. This gives confidence.

Several other hints to successful speaking in public are:

(a) Number the sheets of your notes clearly so that you do not confuse the order in which you want to say certain things.

(b) Speak distinctly, so that a person at the back of the room can hear as clearly as one at the front.

(c) If your mouth feels dry before you start, and there is no glass of water handy, relax your lower jaw, letting your lips scarcely touch each other, for a few moments. You will feel your mouth watering and the dryness will disappear.

(d) Remember the value of the pause in letting an important item of information sink in. Don't rush from sentence to sentence fearing that a pause means you have run dry. Never speak too quickly, and don't use big words if smaller ones will suit the purpose.

(e) Avoid mannerisms and poses. Just stand up straight with your shoulders back. Don't stick your hands in your pockets. Try to forget them and yourself as soon as you can. Don't wander about restlessly, but move easily.

(f) Use statistics carefully. Cut them down to a minimum because they rarely register when merely spoken. Put them in a hand-out or on a slide, and let them make a visual impact.

(g) Treat your audience as human beings. Talk *to* them rather than *at* them, and don't talk to the wall at the back of the room, or the window, ignoring your audience.

These points may help a little, but the main guides to follow in delivering an oral report remain:

(a) Know thoroughly what you are talking about.

(b) Prepare your material carefully and speak with the aid of notes.

(c) Finally, practice what you are going to say at least once privately beforehand.

4000 words

COMPREHENSION TEST

Summary

In about 250 of your own words, summarize the main points of this report. Do **not** *refer back to the passage.*

Discussion—discuss one of these questions (orally if in a group, in writing if studying alone). *You may refer back to the passage.*

1. How important is effective report writing to business and industry?

2. Is there such a thing as "business" or "technical" English?

3. Should schools and colleges train people to write reports, or is this a job for firms themselves?

4. How important is effective writing to efficient reading?

5. How does the science of ergonomics contribute to the provision of materials that can be read efficiently?

Note

You may enter your reading speed on the progress graph and may also, if you wish, enter an estimate of your comprehension level on this passage.

PRACTICE FOR THE COMING WEEK

Select a number of pieces of reading matter that are typical of the kind of material with which you have to deal at work. Try to identify the specific problems you encounter in reading them. Practice reading systematically and purposefully along the lines so far suggested.

Before you work through the next section in this book, write a brief assessment of how far and in which ways your ability has improved since you began this course of reading skills improvement. Limit yourself to five to ten lines.

If you require additional practice, continue with some of the practice suggestions made in earlier sections.

CHAPTER SUMMARY

1. **Some Problems in Reading**
   a) Newspapers should be previewed first. Read more than one if possible. Avoid the "popular" press.
   b) Correspondence should also be skimmed first. Frequently, only letters requiring personal action need to be read in full.

   c) In journals and magazines, make good use of any summaries provided. Read a few of the opening and closing paragraphs and skim the rest for major points. Note the advertising layout.

   d) In reports, read the summary first, then the table of contents. Next, the report as a whole should be skimmed and the sections for special attention noted. Now the report can be read, if this is necessary in the light of the information already revealed.

   e) Do a certain amount of reading for pleasure. The more the better, because this promotes the improvement of quality of comprehension.

2. **Vocabulary**
   a) Always have a dictonary handy when reading.
   b) Undertake systematic vocabulary building.

3. **Anticipation**
   a) Try to foresee the end toward which the writer is heading.
   b) Active involvement in reading encourages better anticipation.

4. **Concentration**
   a) Avoid regressions, define purpose, and anticipate well to improve concentration.
   b) Active involvement in reading helps concentration. Do not be a passive, inert, let-the-author-lead reader. Know what you want to know and go out looking for it.

SIX

Final Assessment

FINAL TESTS

You have now completed this short course of training in the improvement of reading skills. The tests in this section will enable you to see how far you have progressed since you began the course some weeks ago.

Altogether, these tests will take you approximately one hour to complete. Select a time and place that will enable you to attempt them without interruption. Try to pick the same place and the same time of day at which you attempted the Initial Tests. In this way, you will be able to make a more accurate comparison with your reading comprehension at the beginning of the course.

At this point, spend a few seconds previewing the tests, noting the subject matter and the length of each and the location of the test questions. Just glance through this Final Test section quickly and do not try to read anything in detail. Remember that all you require is an idea of the size of the task that faces you.

1. *Look quickly through the Final Test section.*

2. *Make sure that you have with you your watch, your notebook, and a pen and pencil.* Preferably, sit at a desk or table and place these materials within easy reach.

3. *You are now ready to begin.* Remember the procedure for reading each passage:

a) Begin timing and begin reading.

b) Read the passage once only as quickly as you feel you can without loss of comprehension.

c) Answer sections A and B of the Comprehension Test (there is no section C this time).

d) When you have done this, calculate your speed in "words per minute" (Table on page 254).

e) Check your answers to the Test against those given on page 260.

f) Enter your results on the Progress Graphs.

g) Turn to the next passage and follow the same procedure.

Begin NOW.

Instructions

Read through the passage once only *as quickly as you can without loss of comprehension.*

Begin timing and begin reading NOW.

NINETEENTH CENTURY SCIENCE—I
by J. G. Bruton

Advances in the field of science in the nineteenth century produced three important scientific theories—the conservation of energy, the conservation of matter and evolution. The first two pointed towards materialism; the third produced a revolution in thought similar to that which occured in the seventeenth century.

The increase in knowledge was so great that science began to split up into the sub-branches we know today. Specialisation in rather narrow fields became more and more necessary.

An important characteristic of intellectual life in the nineteenth century was a growing respect for science. Comte (1798-1857), a French philosopher, evolved a system called "positivism" in which science finally took the place of theology and metaphysics.

Attempts were made to make use of the scientific method in almost all branches of thought; to treat history, for example, as a science rather than an art.

For the first time, men of science came into open conflict with the views of philosophers. The leading philosopher of the beginning of the nineteenth century, Hegel (1770-1831), knew nothing of science. His view was that everything in the universe, including matter, was essentially moral or spiritual. His philosophy was reasonable when applied to history and morals, but seemed absurd to men of science, when applied to the natural sciences.

The great scientific principles of the indestructibility of matter and of the conservation of energy led to the view that the essential reality of the universe was matter and that the behaviour of matter obeyed scientific laws. Organic matter, the chemists showed, obeyed laws just as inorganic matter did, and life itself, according to the theory of evolution, had developed mechanistically.

John Tyndall, the physicist, stated that just as atoms and molecules combined to produce the beautiful and complicated forms of crystals, so they also form the more complicated living matter, plants and animals. Thoughts are the result of chemical activity in the brain. If the brain is damaged, the mind cannot function normally. So the mind is a product of matter. Tyndall also stated publicly that what had prevented the advance of science until comparatively recent times had been the opposition of the Church.

The conflict between science and religion went on all through the later part of the nineteenth century. While people like W. G. Ward believed that knowledge is gained by intuition and revelation, and believed in the supernatural, others like Thomas Huxley held that knowledge could be gained only by experience and the scientific method and did not accept the supernatural. Huxley invented the word "agnostic" to describe his own attitude to religion.

The attitude of the Roman Catholic Church during the conflict was unmoved. As a result of the dogma of the infallibility of the Pope, made public in 1870, there could be no argument about his utterances or matters of faith and morals. Catholics who were guilty of "modernism" were excommunicated. Among Protestants there was a group of fundamentalists who believed in the absolute truth of the Bible story of the creation and refused to accept evolution.

500 words

Write down the time taken to read this passage and then attempt the Comprehension Test.

COMPREHENSION TEST

Select the most suitable answer in each case.
*Do **not** refer back to the passage.*

A. Retention

1. An important characteristic of intellectual life in the nineteenth century was:
   a) a love of literature.
   b) a growing respect for science.
   c) a respect for metaphysics.
   d) a belief in God.

2. The philosophical system called "positivism" was evolved by:
   a) Hegel.
   b) Tyndall.
   c) The Pope.
   d) Comte.

3. Hegel's view was that everything in the universe was:
   a) based on science.
   b) essentially moral or spiritual.
   c) essentially religious.
   d) the result of accident.

4. John Tyndall believed that thoughts were the result of:
   a) electrical activity in the brain.
   b) the movement of atoms.
   c) chemical activity in the brain.
   d) blood circulation in the brain.

5. The dogma of the infallibility of the Pope was made public in:
   a) 1798.
   b) 1831.
   c) 1857.
   d) 1870.

B. Interpretation

6. The theory of the conservation of energy held that:
   a) the total quantity of energy in the universe was invariable.
   b) energy could be stored up for future use.
   c) the universe came into being as a result of natural causes.
   d) the dogma of the infallibility of the Pope was incorrect.

7. It would be reasonable to suppose that "positivists" were:
   a) Christians.
   b) deists.
   c) Catholics.
   d) atheists.

8. The important scientific theories of the nineteenth century:
   a) confirmed the existence of God.
   b) challenged the existence of God.
   c) converted everyone to Catholicism.
   d) turned everyone against religion.

9. An agnostic is one who:
    a) believes in God.
    b) sometimes believes in God.
    c) doubts the existence of God.
    d) denies the existence of God.

10. In the nineteenth century:
    a) no Catholics were scientists.
    b) few Catholics were scientists.
    c) many Catholics were scientists.
    d) all Catholics were scientists.

Convert the time taken to read the passage into "words per minute" by using the Reading Speed Conversion Table on page 254.

Enter the result on the Progress Graph on page 256.

Check your answers to the Comprehension Test against the answers given on page 260. Enter the result on the Progress Graph on page 256.

Proceed to the next reading exercise.

Instructions

Read through the passage once only *as quickly as you can without loss of comprehension.*

Begin timing and begin reading NOW.

NINETEENTH CENTURY SCIENCE—II

by J. G. Bruton

Newton and his contemporaries believed that God created the universe in essentially the same form as it has today. The nineteenth century view was very different. Laplace put forward the view that the solar system began as a glowing, rotating gas from which the planets condensed and cooled: Hutton and Lyell described the gradual transformation of the surface of the earth during millions of years, how mountains had risen and been worn away, how continents had sunk beneath the sea and risen again; and Darwin offered a picture of the appearance of primitive forms of life which developed during vast periods of geological time into an enormous variety of complex plants and animals and finally into man.

Herbert Spencer (1820-1903), an influential but not a great philosopher, tried to apply the theory of evolution to sociology, psychology and ethics, but without great success. The theory of natural selection was put forward to explain the evolution of living things and cannot be applied to astronomical or geological evolution. The word "evolution," when used outside the field of biology, can mean little more than "change," since there is no principle that can be discovered at work in other fields.

Although progress in science in the nineteenth century was very considerable, progress in technology was even greater. The immense development of industry in Europe and the USA produced a rise in the average standard of living greater than that achieved in the two thousand years before. The population of Europe, which had not increased greatly since 1700, rose from 187 millions in 1800 to 400 millions in 1900.

The industrial revolution, begun in Britain in the eighteenth century, continued at increasing speed and spread to other European countries. A method for manufacturing cheap steel was invented by Sir Henry Bessemer in 1856. British coal production rose from 10 million tons in 1800 to 60 million in 1850 and 210 million in 1900.

The steam engine, which helped to make such progress possible, was invented by men who owed little to theoretical science. Although science had little influence on industrial progress in the first half of the nineteenth century, it made possible the creation of entirely new industries in the second half.

Railways and steamships made great changes possible in systems of transport. Railways were built to satisfy the need of industry for a means of transporting raw materials and manufactured goods. The existence of railways in turn helped in the development of industry. By means of railways it was possible for the people of the eastern part of the USA and for the large numbers of people arriving from Europe to spread across the American Continent.

Science made possible the establishment of two entirely new industries—the electrical and the fine chemical industries. The first London telephone exchange was set up in 1879 and the first station for supplying electricity to private users began to operate in New York, in 1882. From 1870 a lot of research was carried out on synthetic dyes, mainly in Germany, and many new ones were discovered.

510 words

Write down the time taken to read this passage and ther attempt the Comprehension Test.

COMPREHENSION TEST

Select the most suitable answer in each case.
Do not refer back to the passage.

A. Retention

1. Laplace put forward the view that the solar system began:
   a) as the Garden of Eden.
   b) as it is today.
   c) as a number of dead planets.
   d) as a glowing, rotating gas.

2. Spencer applied the theory of evolution to sociology:
   a) without great success.
   b) with some success.
   c) with reasonable success.
   d) with great success.

3. The word "evolution," when used outside the field of biology, can mean little more than:
   a) turning around.
   b) change.

    c) improvement.
    d) progress.

4. In the nineteenth century, progress in science was:
    a) very considerable.
    b) reasonable.
    c) small.
    d) negligible.

5. The population of Europe in 1900 was:
    a) 200 million.
    b) 300 million.
    c) 400 million.
    d) 500 million.

B. Interpretation

6. The influence of science during the nineteenth century:
    a) declined.
    b) increased.
    c) remained the same.
    d) began.

7. The importance of railways in the development of the USA was:
    a) considerable.
    b) reasonable.
    c) small.
    d) negligible.

8. Darwin's theory of evolution was:
    a) universally applicable.
    b) only applicable to man.
    c) only applicable to science.
    d) of little application.

9. The progress in technology and population increases were:
    a) unrelated.
    b) closely linked.
    c) mutually dependent.
    d) unconnected.

10. Technological progress can be said to have begun in:
    a) USA.
    b) Britain.
    c) France.
    d) Germany.

Convert the time taken to read the passage into "words per minute" by using the Reading Speed Conversion Table on page 254.

Enter the result on the Progress Graph on page 256.

Check your answers to the Comprehension Test against the answers given on page 260. Enter the result on the Progress Graph on page 256.

Proceed to the next reading exercise.

Instructions

Read through the passage once only *as quickly as you can without loss of comprehension.*

Begin timing and begin reading NOW.

NINETEENTH CENTURY SCIENCE—III

by J. G. Bruton

The Germans, because they led the world in theoretical chemistry, were the leaders in the fine chemical industry, producing dyes, perfumes, drugs and explosives. The First World War obliged other countries to set up their own chemical industries, with the result that production in Britain and the USA later became greater than that of Germany.

Germany led too in the optical industry, again because of her superior science and because she set up research laboratories for industry. She produced the finest cameras, microscopes, spectrometers and other scientific instruments.

Towards the end of the century industrialised Europe could no longer produce enough food to feed its greatly increased population. Food was imported from North America, where machines like the combine harvest were invented to help to meet the demand for greater production. Refrigeration and canning enabled meat, fish and fruit to be transported to Europe over the oceans of the world. The first cargo of frozen meat reached England in 1880.

The nineteenth century was a period of great public optimism based on enormous scientific and technological advances and rising standards of living. People believed in automatic and inevitable progress.

The origin of the idea of progress goes back about 300 years, to the time of Francis Bacon and the beginnings of modern science. It was an unknown idea in classical times. In the Middle Ages men looked backward to a Golden Age. In the eighteenth century, however, the idea of the progress of knowledge, the result of the development of reason and science, became very general.

The theory of evolution seemed to support the idea of inevitable progress. Everything is evolving, said Spencer, and the direction of evolution must be good. Spencer was convinced that mankind was becoming better and better and would finally reach perfection.

In spite of the horrors of the two world wars, which showed that modern man is still capable of the greatest cruelty, many people still believe in the idea of progress. But it is a belief in purely material progress, the product of technological advance.

Most scientific advances of the nineteenth century were made in France, Britain and Germany. The science of these three countries had some distinctive characteristics. French science was often perfect in form and thought; English science was individualistic and often highly original; German science was thorough and very well organised.

For the first quarter of the century Paris was the scientific centre of the world. In 1794 the Ecole Polytechnique and the Ecole Normale Supérieure were founded, and these trained many of the finest French scientists throughout the century. In 1795, the Academy of Sciences, closed during the Revolution, was re-opened. Although later in the century France produced an outstanding figure,

Pasteur, she never again was as great a leader in science as she had been in the past.

From about 1830 until 1865 the most important scientific developments came from Britain, in the work of Lyell, Faraday, Joule, Darwin and Maxwell. Of these five men Joule, Darwin and Lyell were amateurs with private incomes and Faraday worked at the Royal Institution.

510 words

Write down the time taken to read this passage and then attempt the Comprehension Test.

COMPREHENSION TEST

Select the most suitable answer in each case.

Do not *refer back to the passage.*

A. Retention

1. The leaders in the fine chemical industry in the nineteenth century were:
   a) the British.
   b) the French.
   c) the Germans.
   d) the Americans.

2. The leaders in the optical industry were:
   a) the British.
   b) the French.
   c) the Germans.
   d) the Americans.

3. In the nineteenth century, people believed progress was:
   a) unlikely.
   b) possible.
   c) likely.
   d) inevitable.

4. Spencer was convinced that mankind:
   a) was becoming better and better.
   b) was getting worse and worse.
   c) was perfect.
   d) was doomed.

5. From 1830 to 1865, the most important scientific developments came from:
   a) Britain.
   b) France.
   c) Germany.
   d) USA.

B. Interpretation

6. The First World War obliged several countries to set up their own chemical industries because of:
   a) the need for paint.
   b) the need for explosives.
   c) the need for perfumes.
   d) the need for food.

7. The belief in the perfectability of man suffered a setback as a result of:
   a) the theory of evolution.
   b) the World Wars.
   c) the progress of science.
   d) the ideas of Spencer.

8. Inventiveness in science was a distinctive characteristic in:
   a) USA.
   b) Germany.
   c) France.
   d) Britain.

9. In becoming the scientific center of the world in the first quarter of the nineteenth century, Paris was helped by:
   a) its educational facilities.
   b) the Revolution.
   c) the competition from Germany.
   d) Britain.

10. Most of the important nineteenth-century British
    scientists were:
    a) French.
    b) poor.
    c) engineers.
    d) amateurs.

Convert the time taken to read the passage into "words per minute" by using the Reading Speed Conversion Table on page 254. Enter the result on the Progress Graph on page 256.

Check your answers to the Comprehension Test against the answers given on page 260. Enter the result on the Progress Graph on page 256.

Now, calculate your average score by adding the three results together and dividing by three. Do this for both reading speed and comprehension.

Enter the results on the Progress Graphs on page 256 and regard this as your final score and finishing point for the course.

ANOTHER QUESTIONNAIRE

Now, answer the following short questionnaire, which will enable you to assess other, less tangible benefits you have gained from this course of training.

1. Has your speed of reading improved as much as you had been led to expect that it would?

2. Has your quality of reading comprehension improved or remained the same (ignore apparent gains or losses of less than 10 percent)?

3. Are you pleased with your achievements?

4. Did you do your best or do you feel you could have done better by, for example, practicing more regularly?

5. Did you, in fact, practice regularly and generally carry out the advice and instructions you were given during the course?

6. Do you enjoy reading more now or get more satisfaction from it than you did at the beginning of this course?

7. Do you do more general reading now than you did at the beginning of this course?

8. Do you now find that you can read all that you *have* to read in the course of a week?

9. Are you prepared to continue practicing reading better and faster for another month or two?

10. Do you now remember more of what you have read after reading, for example, a newspaper article?

11. As you read, do you often regress now?

12. Has your concentration when reading improved?

13. Are you now surer of your purpose in reading every piece of written material you encounter?

14. Are you interested in a wider range of subjects and activities now than you were at the beginning of the course?

15. Can you skim effectively now?

16. Do you feel more relaxed now when reading "against the clock"?

17. Are you a more critical reader now?

18. Are you now able to adjust your speed of reading to changes in purpose or the nature and level of difficulty of the material?

19. Can you study effectively now?

20. Have you noticed any improvements in the use of other language communication skills—thinking, listening, speaking, and writing?

ANSWERS TO THE QUESTIONNAIRE

Compare your answers with the comments below on the answers to these questions.

1. *The average reader,* as has already been indicated, *will have increased his speed of reading by about 80 percent.* If your increase was less than 25 percent then this course has probably failed to benefit you very much, but see how the rest of your answers measure up before you draw firm conclusions.

2. *The average reader will have improved slightly, but perhaps not as much as 10 percent in this short course.* A gain of at least 10 percent is necessary for the improvement to be significant. If you have suffered a loss of more than 10 percent, you have probably tried to gain too much too quickly in your speed of reading. Or you may be one of the few people for whom increases in reading speed are not possible.

3. Only you can answer this question, but if you are not satisfied you should try to be particularly honest about your answers to questions 4 and 5.

4. Any training depends for its success on the full cooperation of the student.

5. *Regular practice,* as you may now appreciate, *is essential for permanent success.*

6. Most students find reading a more pleasurable and satisfying experience after a course of this kind. Usually the acquisition of greater skill in any activity gives greater satisfaction.

7. If you do not do more general reading now than you used to, you should do, in the interests of long-term improvements in quality of reading comprehension.

8. *By more skillfully combining your reading techniques, you should now be able to complete all your compulsory reading within the time available.*

9. Unless you are prepared to regard your achievements so far as a beginning only and to continue practicing reading faster and better, you may find that you lose some of the improvement you have made.

10. Many readers will notice a useful improvement in their ability to retain information once read.

11. Regressions should be infrequent and should occur only when necessary to maintain quality of reading comprehension.

12. You may not have noticed any improvement in concentration, but if you are now a better, as well as a faster, reader, this will usually mean that you are making better use of your powers of concentration.

13. The answer to this question must be a definite "Yes" if you are to improve your general efficiency in reading.

14. The passages in this book should have aroused new reading interests. Remember that wide interests are one of the chief characteristics of the efficient reader.

15. *Skimming is a technique that requires practice to achieve maximum effectiveness.* It is easier to develop skill in skimming if you can read at speeds ranging from 300 to 800 w.p.m., but the technique of previewing before reading helps to improve skimming.

16. If you still feel tense when reading under pressure, this indicates that you still lack a certain amount of confidence and that you are probably more concerned with having too much to read in too little time than with efficiently assimilating the meaning and significance of what you are reading. Remember that you must give your complete attention to your reading.

17. You should be a more critical reader now, but refer back to Chapter 3 occasionally when you need to be particularly methodical in approaching reading matter critically.

18. You should be a more **flexible** reader than you were a few weeks ago. Keep asking yourself if you are using the most appropriate reading technique for each piece of reading matter.

19. If you have any particularly important reading to do, remember that you are, in fact, saving time and effort in the long run if you study the material methodically in the first instance.

20. Even though you may not have noticed any improvements yet, you will find that, in the long term, general improvements begin to show themselves.

If your answers to any of these questions are not in line with the answers suggested here, you should revise those chapters in which the particular points are dealt with.

CONTINUATION

Obviously there is no point in taking the trouble to improve the speed and quality of your reading comprehension if you are to find, three or six months later, that the gains you achieved have disappeared. Some attention must therefore be paid to methods of maintaining improvements.

The passages in this section are designed to test your speed and quality of reading comprehension at three- and six-month intervals after you have completed the course. *You should make appropriate entries in your diary to remind you when to take the tests.* If you do take these *follow-up tests,* you will know whether or not the course has been of long-term value to you.

The presence of these tests will, in itself, keep the idea of reading efficiently in the forefront of your mind, which will be an encouragement to you to continue practicing what you have learned. However, *there are three other ways in which you can ensure that your improvements become permanent.*

1. You should remember what we have said about the value of *wide, varied reading.* Set aside a certain amount of time each week for general reading. Read relatively easy and more difficult material in more or less equal amounts and read a mixture of fiction and nonfiction. It might be a good idea to concentrate on reading about subjects you are interested in but have never felt you had the time to follow up. You should also try to read the novels, plays, and even poetry you have always wanted to read, or that have been recommended to you, but for which you have never had enough time.

2. You should make *periodic checks on your speed and quality of reading comprehension* by working through some of the less well-remembered passages in this book again. Follow the same procedure for reading the passages that you followed when you worked through the book the first time. You may find that this is a more useful exercise than practicing on material you have selected yourself but on which no comprehension test has been set, though there is the disadvantage that you may still be familiar with the contents of each passage.

3. You should, *periodically, revise your familiarity with the points made in this book.* It will normally be sufficient simply to skim through each chapter, though if you require to refresh your memory on specific points a more detailed reading should be profitable.

If you follow these methods for maintaining the improvements you have made, you will find that not only do you fare better on the follow-up tests in this section but that you also feel much more confident that the gains are being maintained in your normal, everyday reading at school or work. You may even find that your speed and quality of reading comprehension improve further. This should be particularly true for the quality of your reading comprehension. The final result, in any event, will be that you continue to become a faster and a better reader and generally find yourself dealing more efficiently with your reading. In other words, *there can be no really* **final** *assessment as far as reading improvement is concerned.* Like education, it is a continuing, lifelong process, and you may continue to improve for some considerable time to come if you follow the advice given here.

FOLLOW-UP TESTS (After Three Months)

Three months after working through this book, read through these three passages. They are comparable in standard and content with the Initial and Final Tests.

Instructions

Read through the passage once only *as quickly as you can without loss of comprehension.*
Begin timing and begin reading NOW.

EIGHTEENTH CENTURY SCIENCE—I
by J. G. Bruton

The eighteenth century was an age of great philosophers. From 1690 onwards a group of philosophers—Locke, Berkeley, Hume and Kant—tried to fit science and philosophy together.

John Locke (1632-1704) was a friend of Boyle and Newton, a member of the Royal Society and an experimenter in medicine. His *Essay Concerning Human Understanding,* published in 1690, was the beginning of a long philosophical argument. The origin of Locke's investigations was a discussion with some friends in which they decided that to find any real solutions to fundamental problems it was necessary to study human abilities and to find out what the human mind was capable of. Locke examined his own mind and came to the conclusion that the only origin of knowledge is experience. Locke founded the school of philosophy known as "empiricism"; Descartes had been a "rationalist," believing the origin of knowledge was reason, and that we form ideas and beliefs by deduction from inborn ideas, rather as in geometry. Locke did not accept the existence of inborn ideas; he believed that ideas were the result of experience.

To Locke, the mind at birth was like a clean sheet of paper. Knowledge was obtained through the senses and this knowledge was remembered, ordered and organised. Locke divided ideas into two types—ideas of sensation and ideas of reflection. The mind by reflection organised the simple ideas of sensation into more complex forms. For example, sensations of hardness, smoothness, brownness and squareness could be organised into the idea of a table.

Locke used in psychology an atomic theory like that of physical science. For him sensations were "atoms" introduced by the external senses into the mind, where they were combined into "compounds" by reflection.

Locke accepted the existence of external material objects, such as tables, since he accepted the necessity of something able to produce our ideas of sensation. He took from physics the idea of a distinction between primary and secondary qualities. He believed that the primary qualities of size, shape, motion and number were objective, but that the secondary qualities of colour, taste and smell were entirely subjective; for him they were ideas in the mind, which differed from person to person and under different conditions. It was possible to have objective knowledge of the shape, mass and state of motion of a table, but not of its colour or smell.

Many people of his time accepted Locke's philosophy because of its scientific character, but Bishop Berkeley (1685-1753) did not. Berkeley fought against the weakening of religious belief caused by the success of science. He believed that knowledge of reality came from religious and moral experience and not from the senses and by the scientific method. He tried to show that even the primary qualities of Locke were subjective, and ideas in the mind. A table from one position might appear round, and from another, oval. Its size to the eye depended on distance. He developed this argument to prove that nothing could exist without a mind.

500 words

Write down the time taken to read this passage and then attempt the Comprehension Test.

COMPREHENSION TEST

Select the most suitable answer in each case.
*Do **not** refer back to the passage.*

A. Retention

1. The eighteenth century, the writer says, was an age of:
   a) great scientists.
   b) great politicians.
   c) great writers.
   d) great philosophers.

2. The author of *Essay Concerning Human Understanding* was:
   a) Isaac Newton.
   b) John Locke.
   c) Bishop Berkeley.
   d) David Hume.

3. The origin of Locke's investigations was:
   a) a discussion with some friends.
   b) his friendship with Boyle.
   c) his membership of the Royal Society.
   d) a casual acquaintance.

4. Locke believed that the mind at birth was:
   a) filled with inborn ideas.
   b) filled with foolish ideas.
   c) like a clean sheet of paper.
   d) like a seedling ready for growth.

5. Bishop Berkeley fought against:
   a) the growth of science.
   b) immorality.
   c) repressive politicians.
   d) the weakening of religious belief.

B. Interpretation

6. Locke based his rejection of the existence of inborn ideas on:
   a) personal experience.
   b) scientific experiments.
   c) the existence of God.
   d) a long philosophical argument.

7. The influence of *Essay Concerning Human Understanding* was:
   a) negligible.
   b) slight.
   c) considerable.
   d) greater than any other work of philosophy.

8. The empiricist philosophers believed that:
   a) ideas were the result of experience.
   b) we form ideas and beliefs purely by deduction.
   c) religion was unnecessary.
   d) the origin of knowledge was reason.

9. Locke believed that secondary qualities of color, taste, and smell were entirely subjective because:
   a) they were not as important as primary qualities.
   b) they were not external material objects.
   c) they were never constant under all conditions.
   d) they did not really exist.

10. Bishop Berkeley believed that:
    a) the success of science benefited religion.
    b) science and religion dealt with different but sympathetic matters.
    c) science and religion did not conflict.
    d) there was a conflict between science and religion.

Convert the time taken to read the passage into "words per minute" by using the Reading Speed Conversion Table on page 254. Enter the result on the Progress Graph on page 256.

Check your answers to the Comprehension Test against the

answers given on page 260. *Enter the result on the Progress Graph on page* 256.

Proceed to the next reading exercise.

Instructions

Read through the passage **once only** *as quickly as you can without loss of comprehension.*

Begin timing and begin reading NOW.

EIGHTEENTH CENTURY SCIENCE—II

by J. G. Bruton

David Hume (1711-1776) came to the conclusion that all we can be sure of is the existence of our own ideas. He believed that our knowledge of substance, material or spiritual, is the result of the joining together of perceptions. We believe that a table exists because we join together perceptions of hardness, smoothness, brownness and roundness. But there is no necessary connection between these perceptions.

Hume saw that acceptance of the existence of substance depended on the belief that for every effect there must be a cause. He did not accept the existence of causes as "happenings which make other happenings follow." If two happenings take place, one is not necessarily the result of the other. After we have seen two happenings take place together a number of times, we may believe that one is caused by the other, but there is no real final reason for this. We may say that the sun will rise tomorrow or that sugar dissolves in water, but we can imagine this not happening.

If Hume is correct, we cannot use past happenings to decide what will happen in the future; this destroys the foundation of scientific method, in which we build theories on past events.

Immanuel Kant (1724-1804) developed further the philosophy of Locke and Hume. He believed that the mind by its activity added a contribution to what was learned by experience. His book, *The Critique of Pure Reason,* published in 1781, examined the contribution made by the mind in the gaining of knowledge. It brought together empiricism and rationalism. Knowledge, according to Kant, consists of two parts; that which is given by experience, and that which is given by the mind.

His answer to Hume was that our minds invent causality to make it possible for us to give continuity to our discontinuous perceptions. Experience is organised in our minds also by our ideas of space and time.

Kant was an able physicist and one of his objects as a philosopher was to explain how science and mathematics are possible. The operations of mathematics are the result of our way of perceiving space and time. This same method of perception makes it possible to apply mathematics to science.

The eighteenth century was the "Age of Enlightenment," particularly in France, which was at that time the centre of European civilisation. The French men of learning believed that if reason and science were free to express themselves, they would produce a new outlook. They believed in freedom of thought and expression, in reason, in science and in progress. They learned from Descartes the method of systematic doubt, from Bacon the idea of science as a means of social progress and the idea of gathering encyclopedic knowledge, from Newton the method of physical science and from Locke the science of the mind.

Natural laws, similar to the law of gravitation in physics, were looked for in psychology, political science, economics, history and religion.

490 words

Write down the time taken to read this passage and then attempt the Comprehension Test.

COMPREHENSION TEST

Select the most suitable answer in each case.
*Do **not** refer back to the passage.*

A. Retention

1. The philosopher who came to the conclusion that all we can be sure of is the existence of our own ideas was:
   a) Kant.
   b) Hume.
   c) Descartes.
   d) Newton.

2. Hume saw that the acceptance of the existence of substance depended upon the belief that:
   a) past happenings decide what will happen in the future.
   b) reason and science were free to express themselves.
   c) for every effect there must be a cause.
   d) natural laws had to be obeyed.

3. Hume believed that if two events took place one was:
   a) the result of the other.
   b) never the result of the other.
   c) always the result of the other.
   d) not necessarily the result of the other ·

4. Kant was the author of:
   a) *The Critique of Pure Reason.*
   b) *The Age of Enlightenment.*
   c) *The Operations of Mathematics.*
   d) *Essay Concerning Human Understanding.*

5. The French men of learning believed that if reason and science were free to express themselves they would:
   a) produce a new outlook.
   b) be harmful to religion.

     c) restrict human progress.
     d) destroy natural laws.

B. Interpretation

6. Our perceptions are:
   a) what we experience through our senses.
   b) guesses about the world around us.
   c) what we think.
   d) our own ideas.

7. Hume believed that the fact that some events, such as the sun's rising, were predictable:
   a) demonstrated the logic of cause and effect.
   b) was purely coincidental.
   c) was no final proof of a relationship between cause and effect.
   d) proved the earth was a sphere.

8. Hume's ideas were of the greatest interest to scientists because:
   a) he concluded that all we can be sure of is the existence of our own ideas.
   b) they explained why sugar dissolves in water.
   c) he explained his ideas clearly and concisely.
   d) they were a challenge to the basis of scientific method.

9. Kant believed that:
   a) only the activity of the mind was important.
   b) the mind and experience were both important.
   c) only experience was important.
   d) experience was not important.

10. French men of learning were:
   a) only interested in religion.
   b) only interested in the arts.
   c) not interested in progress.
   d) interested in many branches of science.

Convert the time taken to read the passage into "words per minute" by using the Reading Speed Conversion Table on page 254. Enter the result on the Progress Graph on page 256.

*Check your answers to the Comprehension Test against the
answers given on page* 261. *Enter the result on the Progress Graph
on page* 256.

Proceed to the next reading exercise.

Instructions

*Read through the passage once only as quickly as you can
without loss of comprehension.*

Begin timing and begin reading NOW.

EIGHTEENTH CENTURY SCIENCE—III
by J. G. Bruton

"Nothing that exists can be against or outside Nature,"
wrote Diderot: Montesquieu declared that slavery was
against Nature, and Rousseau popularised the idea of the
natural noble savage. It was believed that science could
teach man to behave naturally and to shape his society
intelligently. If society was natural and rational, man's inborn
goodness would develop towards happiness and perfection.
This was very different from the Christian point of view, that
man is born sinful and that our surroundings are of secondary
importance.

The most important contribution to progress of his
group was the *Encyclopedia,* published in 21 volumes
between 1751 and 1765. Its chief editor was Diderot, and
among its contributor were most of the great French writers
and thinkers of the time—Voltaire, Rousseau, Montesquieu,
d'Alembert, Buffon and Turgot. They did not all hold the
same ideas, but most of them were progressive thinkers.
Many of the things they wrote were unfriendly to the
Christian religion, but they tried to disguise this. Even so,
the *Encyclopedia* was forbidden and publicly burned. Public
opinion did not support the action of the police, but the
last 14 volumes had to be published secretly.

Many French intellectuals were deists, that is to say

that they accepted the existence of a God and held a "natural religion": others were atheists who did not admit the existence of God. David Hume was dining one night in the house of Baron d'Holbach, a German scientist living in Paris. Hume said that he did not believe there were any atheists and that he had never met one. "Then you have been unlucky," replied the Baron. "Here you are at the table with seventeen."

Voltaire, the great French writer, was a deist. For him God was the First Cause and the architect of Newton's universe. "Nobody can doubt that a painting is the work of a skillful artist," he wrote: "Could copies possibly be produced by intelligence and the original not?" Yet he was always attacking established religion. He said that he wanted to prove that it needed only one man to destroy it. He ended most of his letters with the phrase "Ecrasons l'infame"—"We must destroy the vile thing." The vile thing was superstition, both in religion and in ordinary life, and its destruction was the principal idea of the French philosophers.

Towards the end of the eighteenth century, there grew up in Germany a school of thought very opposed to the philosophy of the Encyclopedists. This was the school of *Naturphilosophie* (Nature Philosophy), of which the most important figures were the poet Goethe (1749-1832), the philosopher Schelling and the biologist Oken (1779-1851).

Naturphilosophie was part of a wider movement, the Romantic Movement, which began in Germany as a protest against rationalism, giving more importance to feeling than to reason.

470 words

Write down the time taken to read this passage and then attempt the Comprehension Test.

COMPREHENSION TEST

Select the most suitable answer in each case.
Do not refer back to the passage.

A. Retention

1. The idea of the natural, noble savage was popularized by:
   a) Rousseau.
   b) Diderot.
   c) Montesquieu.
   d) Hume.

2. When it was published, the *Encyclopedia* was:
   a) welcomed by everyone.
   b) forbidden and publicly burned.
   c) welcomed by the Church.
   d) censored and abridged.

3. Voltaire was:
   a) a German scientist.
   b) an English philosopher.
   c) a great French writer.
   d) a great Italian architect.

4. The principal aim of the French philosophers was:
   a) the creation of outrageous ideas.
   b) the destruction of science.
   c) the creation of new forms of Christianity.
   d) the destruction of superstition.

5. The Romantic Movement began in:
   a) England.
   b) France.
   c) Italy.
   d) Germany.

B. Interpretation

6. The Encyclopedists had to disguise their attacks on the Christian religion because:
   a) politicians did not wish to offend the King.
   b) they might have been mistaken and made to look foolish.
   c) the authorities would not allow public criticism of religion.
   d) Christians did not understand science.

7. In the eighteenth century, French intellectuals were generally:
   a) not interested in religion.
   b) slightly interested in religion.
   c) very interested in religion.
   d) obsessed with religion.

8. The French philosophers wanted the destruction of superstition because:
   a) it was the enemy of the scientific approach.
   b) it was the same as religion.
   c) they were atheists.
   d) Voltaire had said they "must destroy the vile thing."

9. The Romantic Movement was:
   a) a natural development of the progress of science.
   b) an answer to the problem of superstition.
   c) a movement confined only to Germany.
   d) a reaction against the preoccupation with science.

10. One of Voltaire's principal preoccupations was:
    a) attacking science.
    b) attacking established religion.
    c) attacking anyone who believed in God.
    d) attacking German philosophers.

Convert the time taken to read the passage into "words per minute" by using the Reading Speed Conversion Table on page 254. Enter the result on the Progress Graph on page 256.

Check your answers to the Comprehension Test against the answers given on page 261. Enter the result on the Progress Graph on page 256.

Now, calculate your average score by adding the three results together and dividing by three. Do this for both reading speed and comprehension. Compare your performance now with your performance on the final tests. Have you maintained your improvements? Have you done any timed practice in reading faster in the last three months?

FOLLOW-UP TESTS (After Six Months)

Six months after working through this book, read through these three passages. They are comparable in standard and content with the other passages on the Russian revolution.

If a six-month check is not required, these passages may be used for additional practice during or after the course.

Instructions

Read through the passage once only as quickly as you can without loss of comprehension.

Begin timing and begin reading NOW.

IT BEGAN WITH THE WOMEN—IV

by Edward Crankshaw

While the Duma Committee wrestled with the formation of an official Government to negotiate with the Tsar, the real business of governing, in so far as there was any, was being conducted by the Soviet in the opposite wing of the palace. Kerensky dashed between the two poles, made himself indispensable, and learnt the secrets of both sides.

Early on 15th March Milyukov was ready with his Government list. It was a Government headed by an enlightened aristocrat, Prince Lvov, which had no imaginable connection with what was happening in the streets. The only concession to the revolutionary mood was the inclusion of Kerensky as Minister of Justice, and Kerensky himself thought twice: he was afraid that the Soviet would disown him for cooperating with what in any country but Russia would have been a conservative cabinet, including as it did such bourgeois politicians as Milyukov, the new Foreign Minister, and Guchkov, Minister for War. At least one member was a millionaire.

Nor was it in any real sense a revolutionary Government. It was dedicated to the restoration of order, to forcing Nicholas to abdicate, and to the inauguration of a constitution and the creation of an elected Constituent Assembly. It was determined to pursue the war with Germany and, apart from Kerensky, it wished to retain the monarchy. The battle with the Soviet was now on.

But it was more than the battle with the Soviet; it was a battle with the people. This came out most vividly in Guchkov's resistance to the Soviet's demand that it should control the army. There were many Soviets now, springing up all over the country and in the battle zone. The Petrograd Soviet was supreme; but it had no real control. The troops recognised only the Soviet and the Soviet could not have stopped this even if it had wished to.

The fight to retain the monarchy was kept secret from the Soviet in the first place. When Nicholas was at last run to earth by Guchkov, the new Government's plan was that he should abdicate and surrender the throne to his 13-year-old son, Alexis, with his brother the Grand Duke Michael, acting as Regent. Nicholas had tried to get back from Mogilev to Petrograd but his train had been diverted, and now he was held a virtual prisoner at Pskov, about 150 miles away.

He was calm; but he decided that because the Tsarevitch's haemophilia was incurable the crown should go direct to Michael. It was not to be. When the Soviet heard of the plan there was uproar. The monarchy must go.

In the end it was the Grand Duke who decided. He would not, he said, accept the crown unless and until it was offered to him by a properly elected Constituent Assembly. That was the end of the road. The Lvov Government was now the official Government of Russia. It was called the Provisional Government because its task was to hold the fort until a Constituent Assembly could be elected. In its pristine form it was doomed to almost immediate failure.

520 words

Write down the time taken to read this passage and then attempt the Comprehension Test.

COMPREHENSION TEST

*Do **not** refer back to the passage.*

A. Retention

1. What was the only concession to the revolutionary mood made in Milyukov's government list?

2. On which issue did Kerensky disagree with the Milyukov's government?

3. How old was Nicholas II's son, Alexis?

4. Why could not Nicholas II get back to Petrograd?

5. Under what terms would the Grand Duke Michael accept the crown?

B. Interpretation

6. What is meant by "the real business of governing"?

7. Why did Milyukov's government list have "no imaginable connection with what was happening in the streets"?

8. Why was the fight to retain the monarchy kept secret from the Soviet?

9. Why was Nicholas II prevented from returning to Petrograd?

10. Why did the Lvov government become the official government of Russia?

Convert the time taken to read the passage into "words per minute" by using the Reading Speed Conversion Table on page 254. Enter the result on the Progress Graph on page 256.

Check your answers to the Comprehension Test against the answers given on page 261. Enter the result on the Progress Graph on page 256.

C. **Discussion**—discuss one of these questions (orally if in a group, in writing if studying alone). *You may refer back to the passage.*

11. What does the writer mean when he says that Milyukov's government list "in any country but Russia would have been a Conservative cabinet"?

12. What were the advantages and disadvantages of Kerensky's unique position?

13. What is meant by "The Petrograd Soviet was supreme; but it had no real control"?

14. Why was the Soviet so strongly opposed to the Monarchy?

15. Why was the provisional government doomed to almost immediate failure?

Proceed to the next reading exercise.

Instructions

Read through the passage once only *as quickly as you can without loss of comprehension.*

Begin timing and begin reading NOW.

IT BEGAN WITH THE WOMEN—V

by Edward Crankshaw

Meanwhile, Lenin, in Zurich, was at first wholly sceptical. Nothing, he was sure, had really changed. There had been strikes and demonstrations and even Soviets before; there would be again, perhaps many times, before the day of revolution dawned. The men and women of his generation would probably not live to see that day, he was saying. He was not quite 47.

It almost seemed as though he wanted it that way. For 14 years all his energies had gone into laying down the law about the true nature of a Bolshevik revolution and intriguing, manœuvring, bullying, anathematising. Only those who agreed absolutely with him were regarded as fit workers for the cause; those who were not with him were against him, and during all these years his closest supporters were constantly being excommunicated and then, recanting, welcomed back into the fold. The supreme task was to keep Bolshevism pure and uncorrupted even though, as it sometimes appeared, the solitary pure and uncorrupted Bolshevik was Lenin himself.

Even after the news of the Tsar's abdication reached Zurich he could write to one of his most loyal supporters, Madame Kollontai, in Stockholm, saying that the so-called revolution in Russia was nothing but a capitalist upheaval. On the whole he thought it best that the Provisional Government should outlaw the Social Democrats; if Marxism were made legal in Russia, then Bolsheviks and Mensheviks might be tempted to draw together, and this would be the deadly sin.

Soon, however, even Lenin began to be affected by the vulgar enthusiasm of his fellow exiles who were sure that indeed something new had happened in Russia, and it was their imperative duty to hurry home and take charge. But how? More arguments, endless discussions—all leading to the famous sealed train provided by the German General Staff to convey Lenin and his friends through the heart of Germany in order to let them loose there and make the Provisional Government's confusion worse.

The Germans were disappointed by the evident determination of the Provisional Government to honour its obligations to their Allies and go on defending Russia. Not only the Provisional Government, the Soviet too. And even those Bolsheviks left in Russia, now coming out of hiding and returning from exile in Siberia, were caught up in the mood. Kamenev and Stalin both thought the revolution was a wonderful thing, that all revolutionaries, no matter how much they had disagreed in the past, must stand together and defend the revolution not only from reaction at home but from the German Kaiser. They wanted peace, but "peace with honour."

Lenin was not interested in honour. He had made this clear enough for the Germans to see. They did not expect him to return to Petrograd and assume command, but they were fairly sure that he could multiply confusion and weaken the Russian will to fight.

470 words

Write down the time taken to read this passage and then attempt the Comprehension Test.

COMPREHENSION TEST

Do **not** *refer back to the passage.*

A. Retention

1. Whom did Lenin regard as "fit workers for the cause"?

2. What was "the supreme task," according to Lenin?

3. How did Lenin and his friends travel to Russia?

4. By whom was the transport provided?

5. What was the intention in assisting Lenin and the revolutionaries to return to Russia?

B. Interpretation

6. Why was Lenin skeptical at first?

7. Why did Lenin think it would be a sin if Bolsheviks and Mensheviks were tempted to draw together?

8. Why did the Germans think the Bolsheviks would cause greater confusion in Russia?

9. What is meant by "peace with honour"?

10. Why did the Germans not expect Lenin to assume command in Petrograd?

Convert the time taken to read the passage into "words per minute" by using the Reading Speed Conversion Table on page 254. Enter the result on the Progress Graph on page 256.

Check your answers to the Comprehension Test against the answers given on page 261. Enter the result on the Progress Graph on page 256.

C. **Discussion**—discuss one of these questions (orally if in a group, in writing if studying alone). *You may refer back to the passage.*

11. What is meant by "He almost seemed as though he wanted it that way" (Para. 2)?

12. Why was it so important "to keep Bolshevism pure and uncorrupted"?

13. What made Lenin think the "so-called revolution in Russia was nothing but a capitalist upheaval"?

14. Why was the train sealed?

15. Why was Lenin not interested in honor?

Proceed to the next reading exercise.

Instructions

Read through the passage **once only** *as quickly as you can without loss of comprehension.*

Begin timing and begin reading NOW.

IT BEGAN WITH THE WOMEN—VI
by Edward Crankshaw

Lenin was throwing off manifesto after manifesto. There was to be immediate peace with Germany, but civil war to the knife at home. The so-called revolution was nothing but a conspiracy between the Russian capitalists and the French and British to overthrow the Tsar in order to prevent him from making a separate peace with Germany. The proletariat must rally round the Soviet and cast down the Provisional Government, and not only the Provisional Government, but all other revolutionary parties, above all the Mensheviks (themselves Marxists to a man).

This distortion of the situation, as set down in Lenin's *Letters from Afar,* is the key to everything that was going to happen—not only in the summer or autumn of 1917, but for decades to come. "Our tactics!" he screamed across Europe in a telegram to a group of Bolsheviks in Norway. "Our tactics: absolute distrust: no support for the new Government: Kerensky especially suspect: arming of proletariat the only guarantee: immediate election Petrograd Duma: no rapprochement with other parties."

All these demands, had they in fact reached the revolutionaries in Petrograd, would have seemed the ravings of a madman. To those who had been swept up in the intoxication of the past days, who had destroyed the monarchy, made a prisoner of the Tsar, begun to organise the striking workers and mutinying soldiers in the streets; who were now engaged in consolidating their revolution, in fighting reaction, in trying to feed the country and in getting the factories working, to bring some order out of chaos, and all the time keep Russia from being overrun by the Germans—to all these even such words of Lenin as did reach them were the words of a man completely out of touch.

How could he know what had happened? How could he understand the real nature of the great explosion which made nonsense of all cherished theories? The workers, the soldier-peasants, had risen and cast off their chains; and now, guided by professional revolutionaries of every colour working shoulder to shoulder, they were going to fashion a new land. Vladimir Ilyich, who had not set foot in Russia for 10 years, would soon see the new situation for himself and help to lead them to final victory.

They were wrong. It was they, Lenin was to tell them, who were out of touch. It began on the train that brought Lenin into the Finland station in Petrograd. Kamenev, Stalin with him, overwhelmed with emotion, boarded the train outside the station and expected elation and congratulations. Lenin was not elated. "What on earth do you think you are up to?" he demanded in effect. "Fraternising with Mensheviks and Socialist Revolutionaries and going on as though you had won a great victory."

460 words

Write down the time taken to read this passage and then attempt the Comprehension Test.

COMPREHENSION TEST

Do **not** *refer back to the passage.*

A. Retention

1. Who, above all, did Lenin say must be cast down?

2. To whom did Lenin send a telegram?

3. How would Lenin's demands have appeared had they reached the revolutionaries in Petrograd?

4. How long had Lenin been away from Russia?

5. In what kind of mood was Lenin when he arrived in Petrograd?

B. Interpretation

6. Why did Lenin want "immediate peace with Germany, but civil war to the knife at home"?

7. Do you feel that the writer approves or disapproves of Lenin's actions?

8. Why was "Kerensky especially suspect"?

9. What is meant by "the intoxication of the past days"?

10. Why did Lenin consider that it was his supporters who were, in fact, out of touch?

Convert the time taken to read the passage into "words per minute" by using the Reading Speed Conversion Table on page 254. Enter the result on the Progress Graph on page 256.

Check your answers to the Comprehension Test against the answers given on page 262. Enter the result on the Progress Graph on page 256.

C. **Discussion**—discuss one of these questions (orally if in a group, in writing if studying alone). *You may refer back to the passage.*

11. "The so-called revolution was nothing but a conspiracy between the Russian capitalists and the French and British to overthrow the Tsar in order to prevent him from making a separate peace with Germany." Do you think there is any evidence to support this view?

12. Examine the first sentence of the second paragraph. How far is the writer's opinion fair?

13. How far was it to Lenin's advantage that he arrived in Russia after most of the action of the revolution itself had taken place?

14. What is meant by "professional revolutionaries" (Para. 4)? Is the term justified?

15. What differences would it have made to the ultimate success of the revolution if Lenin had not been so fanatical in his attitudes?

Now, calculate your average score by adding the three results together and dividing by three. Do this for both reading speed and comprehension. Compare your performance now with your performance on the final tests. Answer the following questions:

1. Have you maintained the increases in reading speed during the last six months?

2. Have you maintained the quality of your reading comprehension over the last six months?

3. Do you feel you are a more purposeful, flexible, and generally efficient reader now?

4. Have you carried out the continuation work recommended?

5. How much of the instruction given in this book can you remember now? Why not work through some of the less well-remembered parts of it again?

SEVEN

Records and Reference

This final chapter of the book contains several sections which will help both teachers and students to use it with greater ease and effectiveness.

READING SPEED CONVERSION TABLE   page 254

This table will enable the reader to calculate his reading speed quickly and effectively.

PROGRESS GRAPHS   page 256

These will enable the reader to translate his results into a form of visual aid which will be more effective in stimulating progress. Both graphs have been divided into 24 sections, corresponding to the 24 passages in the book which test Comprehension. These include the passage on Report Writing, although this is essentially a Summary and Discussion passage.

ANSWERS TO TESTS   page 257

The answers to the "multiple-choice" and the "open-ended" questions in sections A (Retention) and B (Interpretation) of the Comprehension Tests are given in this section.

FURTHER READING   page 263

This section contains further reading suggestions on the subject of reading efficiency, suggestions for teachers, and suggestions for further reading on other aspects of the improvement of communication skills.

NOTES FOR TEACHERS   page 266

A brief guide to the principal points to be borne in mind when organizing reading efficiency training courses in schools, colleges, and businesses.

A TYPICAL PROGRESS GRAPH FOR A RAPID READING COURSE

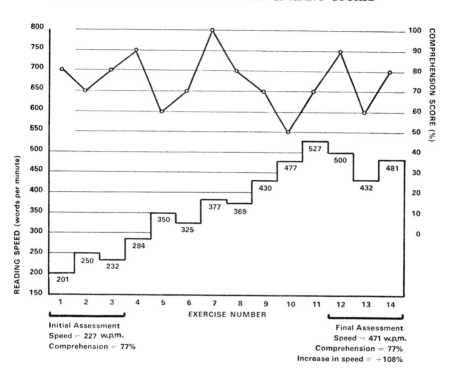

READING SPEED CONVERSION TABLE
Number of Words in Passage

Min/Secs	410	430	460	470	490	500	510	520	800	1200	1300	1600	1700	1800	2000	3200	4000	4500
0.30	820	860	920	940	980	1000	1020	1040	1600	2400	2600	3200	3400	3600	4000	6400	8000	9000
0.35	703	738	790	807	840	858	875	892	1375	2060	2230	2750	2920	3090	3430	5480	6870	7720
0.40	615	646	691	705	735	762	765	780	1200	1800	1950	2400	2550	2700	3000	4800	6000	6750
0.45	547	573	614	627	653	667	680	693	1065	1600	1735	2130	2270	2400	2670	4270	5330	6000
0.50	493	517	553	564	588	600	613	624	960	1440	1560	1925	2040	2160	2400	3840	4800	5400
0.55	447	469	502	513	534	546	557	567	873	1310	1420	1745	1855	1965	2180	3490	4370	4900
1.00	410	430	460	470	490	500	510	520	800	1200	1300	1600	1700	1800	2000	3200	4000	4500
1.05	378	397	425	433	452	462	471	480	738	1110	1200	1475	1570	1660	1850	2950	3690	4150
1.10	352	369	395	403	421	429	438	446	686	1030	1115	1375	1455	1545	1710	2740	3430	3860
1.15	328	344	368	376	396	400	408	416	640	960	1040	1280	1360	1440	1600	2560	3200	3600
1.20	308	323	345	352	367	375	383	390	600	900	975	1200	1275	1350	1500	2400	3000	3390
1.25	289	304	325	332	346	353	360	367	565	848	918	1130	1200	1270	1410	2260	2830	3170
1.30	273	287	307	313	327	333	340	347	534	800	867	1065	1135	1200	1330	2130	2670	3000
1.35	259	272	290	297	309	316	322	329	506	758	821	1010	1075	1135	1260	2020	2530	2840
1.40	246	258	276	282	294	300	306	312	480	721	781	960	1020	1080	1200	1920	2400	2700
1.45	234	246	163	268	280	286	291	297	457	686	743	915	972	1025	1140	1830	2280	2570
1.50	224	235	251	256	267	273	278	284	437	655	710	873	927	982	1090	1745	2180	2460
1.55	214	224	240	245	256	261	266	271	418	627	678	835	887	940	1050	1670	2090	2350
2.00	205	215	230	235	245	250	255	260	400	600	650	800	850	900	1000	1600	2000	2250
2.05	197	206	221	226	235	240	245	249	384	576	624	768	816	863	960	1535	1920	2160
2.10	189	199	212	216	226	231	236	240	369	554	600	739	785	830	925	1475	1845	2080
2.15	182	191	205	209	218	222	227	231	356	534	578	711	756	800	890	1422	1775	2000
2.20	176	185	198	202	210	211	219	223	343	515	557	686	730	773	860	1372	1715	1930
2.25	170	178	191	195	203	207	211	215	331	497	538	663	705	746	825	1325	1655	1865
2.30	164	172	184	188	196	200	204	208	320	481	520	640	680	720	800	1280	1600	1800
2.35	159	167	178	182	190	194	198	201	310	464	503	619	658	697	775	1238	1550	1740
2.40	154	161	172	176	184	188	192	195	300	450	487	600	638	675	750	1200	1500	1690

2.45	1635	1455	1165	725	655	618	582	473	437	291	189	186	182	178	171	167	157	149
2.50	1590	1410	1130	705	636	600	565	459	424	283	184	180	177	173	166	163	152	145
2.55	1545	1370	1100	685	618	583	550	447	412	275	178	175	172	168	161	158	148	141
3.00	1500	1335	1066	667	600	565	533	433	400	267	173	170	167	163	157	153	143	137
3.10	1420	1260	1010	633	568	537	505	411	379	253	164	161	158	155	148	145	136	130
3.20	1350	1200	960	600	541	512	480	391	360	240	156	153	150	147	141	138	129	123
3.30	1285	1145	915	572	515	486	457	372	343	229	149	146	143	140	134	132	129	117
3.40	1225	1090	873	546	492	464	437	355	327	218	142	139	137	134	128	126	117	112
3.50	1175	1045	836	522	470	444	418	340	314	209	136	133	131	128	123	120	112	107
4.00	1125	1000	800	500	450	425	400	325	300	200	130	128	125	122	117	115	108	103
4.10	1080	960	768	480	433	408	384	313	288	192	125	123	120	118	113	110	103	
4.20	1040	925	740	462	416	393	370	300	277	185	120	118	115	113	109	106		
4.30	1000	890	712	444	400	378	356	289	267	178	116	113	111	109	104	102		
4.40	965	860	686	428	386	365	343	279	258	172	111	109	107	105	101			
4.50	932	830	664	413	373	352	332	269	249	166	108	106	104	102				
5.00	900	800	640	400	360	340	320	260	240	160	104	102	100					
5.15	857	760	610	381	343	324	305	248	229	152								
5.30	818	727	582	363	328	309	291	237	218	145								
5.45	783	695	558	348	313	296	279	226	209	139								
6.00	750	667	534	333	300	284	267	217	200	133								
6.30	692	615	492	308	277	262	246	200	185	123								
7.00	643	572	457	286	257	243	229	186	172	114								
7.30	600	535	426	267	240	227	213	173	160	107								
8.00	563	500	400	250	225	213	200	163	150	100								
8.30	530	470	376	235	212	200	188	153	141									
9.00	500	445	356	222	200	188	178	144	133									
9.30	473	422	337	211	189	178	169	137	126									
10.00	450	400	320	200	180	170	160	130	120									
11.00	409	363	290	182	164	155	145	118	109									
12.00	375	333	266	167	150	142	133	108	100									
13.00	346	308	246	154	139	131	123	100										
14.00	321	286	228	143	129	121	114											
15.00	300	267	214	133	120	113	107											

PROGRESS GRAPHS
Comprehension (*Sections A and B of Tests*)

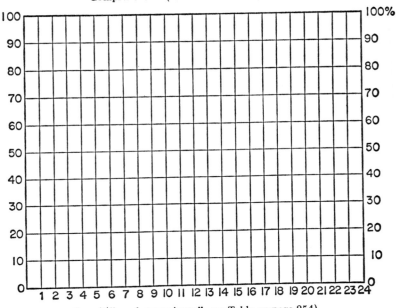

Speed ("words per minute"–see Table on page 254)

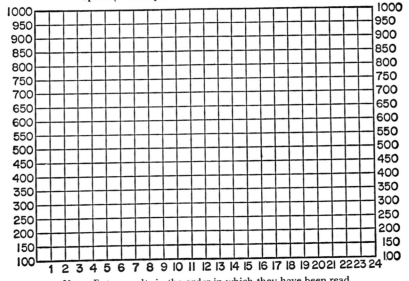

Note: Enter results in the order in which they have been read.

ANSWERS TO COMPREHENSION TESTS

The Troubles of Shopping in Russia (pages 12-17)

1. d.   2. c.   3. b.   4. b.   5. a.   6. d.   7. c.   8. d.
9. b.   10. a.

Seventeenth Century Science—I (pages 18-22)

1. c.   2. b.   3. a.   4. d.   5. b.   6. c.   7. b.   8. a.
9. d.   10. a.

Seventeenth Century Science—II (pages 22-26)

1. b.   2. a.   3. c.   4. a.   5. d.   6. b.   7. d.   8. a.
9. d.   10. a.

Seventeenth Century Science—III (pages 26-30)

1. c.   2. d.   3. a.   4. c.   5. b.   6. a.   7. c.   8. a.
9. d.   10. b.

It Began with the Women—I (pages 48-51)

Note: Where alternative answers are offered, either may be
   awarded a mark.

1. Rasputin.
2. Mogilev.
3. March 1917 (the 8th, in fact).
4. "Give us bread!"

5. Workers and/or soldiers.
6. They were not at all aware.
7. There was no hint of the oppression to come.
8. Nothing was achieved.
9. Estranged/distant.
10. They were not organized at this stage and had no real sense of purpose.

Can Man Survive? (pages 51-56)

1. d.   2. b.   3. c.   4. a.   5. d.   6. a.   7. a.   8. c.
9. b.   10. d.

The Man Who Built Liberia (pages 56-62)

1. c.   2. b.   3. a.   4. c.   5. a.   6. b.   7. a.   8. d.
9. b.   10. b.

Colourful Diversity of Culture (pages 81-87)

1. d.   2. b.   3. b.   4. a.   5. c.   6. a.   7. d.   8. a.
9. b.   10. c.

Sri Lanka's Do-It-Yourself Food Drive (pages 88-92)

1. b.   2. a.   3. c.   4. d.   5. c.   6. b.   7. b.   8. c.
9. d.   10. b.

It Began With the Women—II (pages 93-96)

1. Revolutionary politicians.
2. The Cossacks refused to charge the crowd/the demonstrators were angry, not pleading.

3. They did practically nothing.
4. Dissolve itself.
5. Violent.
6. He was too far away/the Duma would no longer obey him/The Petrograd garrison was on the point of mutiny.
7. They did not know what to do next/Mass movements need leaders if they are to achieve anything more than chaos.
8. Fear of later reprisals from the Tsar/they were unwilling to accept responsibility/they knew nothing of what was going on in the country as a whole.
9. They had locked themselves inside the Tauride Palace.
10. To show their discontent with the oppressive government of the Tsar/to show that they were in control of the situation.

It Began with the Women—III (pages 126-128)

1. Trotsky.
2. Socialist Revolutionaries, Mensheviks, Bolsheviks.
3. Zurich.
4. Tauride Palace.
5. Kerensky.
6. An elected council of workers and soldiers.
7. They recognized only the Soviet as their source of authority.
8. To keep a continuous control over the course of the revolution.
9. He had legal knowledge and skill in persuasion *or* he had contact with both sides.
10. To establish some form of law and order and to try to remedy the immediate cause of revolt/to assert its own authority.

Intelligence: A Changed View (pages 170-176)

1. c.   2. b.   3. a.   4. a.   5. d.   6. b.   7. a.   8. a.
9. c.   10. d.

The Think-Tank Predictors (pages 177-182)

1. a.   2. c.   3. d.   4. b.   5. a.   6. c.   7. a.   8. d.
9. c.   10. a.

Nineteenth Century Science—I (pages 210-214)

1. b.   2. d.   3. b.   4. c.   5. d.   6. a.   7. d.   8. b.
9. c.   10. b.

Nineteenth Century Science—II (pages 214-218)

1. d.   2. a.   3. b.   4. a.   5. c.   6. b.   7. a.   8. d.
9. b.   10. b.

Nineteenth Century Science—III (pages 218-222)

1. c.   2. c.   3. d.   4. a.   5. a.   6. b.   7. b.   8. d.
9. a.   10. d.

Eighteenth Century Science—I (pages 228-231)

1. d.   2. b.   3. a.   4. c.   5. d.   6. a.   7. c.   8. a.
9. c.   10. d.

Eighteenth Century Science—II (pages 232-235)

1. b. 2. c. 3. d. 4. a. 5. a. 6. a. 7. c. 8. d.
9. b. 10. d.

Eighteenth Century Science—III (pages 236-239)

1. a. 2. b. 3. c. 4. d. 5. d. 6. c. 7. c. 8. a.
9. d. 10. b.

It Began with the Women—IV (pages 240-242)

1. The inclusion of Kerensky.
2. The retention of the monarchy.
3. 13.
4. His train had been diverted.
5. That it should be offered by a properly elected constituent assembly.
6. The maintenance of law and order.
7. The only "revolutionary" in it was Kerensky.
8. The Soviet strongly opposed the monarchy.
9. He might have been able to regain power if he had returned.
10. At the time, there was no other body capable of governing because the Soviets were not yet ready to do this.

It Began with the Women—V (pages 243-246)

1. Those who agreed absolutely with him.
2. To keep Bolshevism pure and uncorrupted.

3. In a sealed train.

4. The German General Staff.

5. The Germans hoped they would worsen the confusion in Russia.

6. There had been strikes, demonstrations and Soviets before and nothing had come of them.

7. Bolshevism would not have remained pure and uncorrupted.

8. The Bolsheviks would not accept the provisional government and would try to overthrow it.

9. A negotiated withdrawal from the war that would not appear to be a surrender.

10. Probably because they thought the provisional government was more firmly established and that it would prevent this.

It Began with the Women—VI (pages 246-249)

1. The Mensheviks.

2. A group of Bolsheviks in Norway.

3. As the ravings of a madman.

4. 10 years.

5. Not elated.

6. He wanted to be able to concentrate on achieving complete success for the aims of Bolshevism.

7. Disapproves.

8. He had contact both with the Soviet and the provisional government.

9. The violence and the excitement of the revolution.

10. They had lost sight of the long-term objectives of the revolution in the elation of immediate success.

SUGGESTED FURTHER READING

If, as you have worked through the course in this book, you have become more aware of, and more interested in, problems of increasing reading efficiency and wish to discover what other writers have to say about various aspects of reading, you will find the following books helpful. Teachers will also find the list useful and there is a special section for those who wish to develop their own reading efficiency training courses.

Books on Reading Improvement

Bayley, H. *Quicker Reading.* New York: Pitman, 1957.

Braam, L. S., and Sheldon, W. D. *Developing Efficient Reading.* New York: Oxford University Press, 1959.

*De Leeuw, M. & E. *Read Better, Read Faster.* Baltimore: Penguin Books (Pelican), 1965.

*Fry, E. *Read Faster: A drill book.* New York: Cambridge University Press, 1963.

Gainsbury, J. C. *Advanced Skills in Reading.* New York: Macmillan, 1962.

Gilbert, D. W. *Power and Speed in Reading.* Englewood Cliffs, N. J.: Prentice-Hall, 1956.

Hartman, D. *Speed Reading Course*, National Poetry Association, 1958.

Herrick, M. C. *The Collier Quick and Easy Guide to Rapid Reading.* New York: Collier, 1963.

Leedy, P. D. *Improve Your Reading* (previously entitled *Reading Improvement for Adults*). New York: McGraw-Hill, 1964.

Lewis, N. *How to Read Better and Faster.* New York: Crowell, 1958.

McCorkle, R. B., and Dingus, S. D. *Rapid Reading.* Littlefield, Adams & Co., 1958.

*Mares, C. *Efficient Reading.* London: English Universities Press, 1964.

Miller, L. L. *Increasing Reading Efficiency.* New York: Holt, Rinehart & Winston, 1964.

Miller, L. L. *Maintaining Reading Efficiency.* New York: Holt, Rinehart & Winston, 1967.

*These books are particularly recommended for additional practice.

Pitkin, W. B. *The Art of Rapid Reading.* New York: Grosset & Dunlop, 1929.

Reading Laboratory Inc. *Double Your Reading Speed.* New York: Fawcett, 1964.

Shefter, H. *Faster Reading Self Taught.* New York: Washington Square Press, 1959.

Smith, N. B. *Read Faster.* Englewood Cliffs, N. J.: Prentice-Hall, 1958.

Spache, G. A., and Berg, P. C. *The Art of Efficient Reading.* New York: Macmillan, 1955.

Tinker, M. A. *Basis for Effective Reading.* Minneapolis: University of Minnesota Press, 1965.

*Wainwright, G. R. *Towards Efficiency in Reading.* New York: Cambridge University Press, 1968.

Waldman, J. *Rapid Reading Made Simple.* New York: Doubleday, 1958.

*Webster, O. *Read Well and Remember.* London: Hutchinson, 1965.

Witty, P. *How to Become A Better Reader.* Chicago: Science Research Association, 1953.

Wrenn, C. G. and Cole Ltd. *How to Read Rapidly and Well.* Stamford University Press, 1954.

There are two other books which may be of interest to any teachers of younger children who may be looking for material that is more appropriate for their pupils.

Chapman-Taylor, Y., and Ballard, B. A. *Read and Enjoy.* London: Nelson, 1967. (Nelson also publish a series of Rapid Reading books for practice work by younger children.)

Ridout, R. *English Now: 4.* London: Ginn, 1968.

In addition, newspapers and magazines are good sources for material, but teachers will obviously have to prepare their own comprehension tests for these. One point to remember in the selection of material is that, to facilitate transfer of training, passages should be taken from the kinds of sources that students would normally be encountering. Care should be taken that, particularly in the early stages of a course, this material is not too difficult. More demanding and more varied material may be used when the new techniques are more firmly established.

Books for Teachers

Diack, H. *Reading and the Psychology of Perception.* Nottingham, England: Ray Palmer Ltd (Nottingham).

*These books are particularly recommended for additional practice.

Fry, E. *Teaching Faster Reading*. New York: Cambridge University Press, 1963.

Gates, A. I. *Improvement of Reading*. New York: Collier-Macmillan, 1947.

Harris, A. *How to Increase Reading Ability*. New York: Longman, 1961.

Harris, A. J. (ed.). *Readings on Reading Instruction*. New York: McKay, 1963.

Lefevre, C. A. *Linguistics and the Teaching of Reading*. New York: McGraw-Hill.

Wainwright, G. R. *Efficiency in Reading: A report on courses*. Pembroke Dock, Wales: Dock Leaves Press, 1965.

Books on Related Subjects—A Selection

Thinking:

Chase, S. *Guides to Straight Thinking*. New York: Harper, 1956.

Stebbing, L. S. *Thinking to Some Purpose*. Baltimore: Penguin Books (Pelican).

Thouless, R. *Straight and Crooked Thinking*. London: Pan Books.

Speaking:

Gondin, W. R., and Mammen, E. W. *The Art of Speaking Made Simple*. London: W. H. Allen, 1970.

Henderson, A. M. *Good Speaking*. London: Pan Books, 1956.

Lawrence, R. S. *Public Speaking*. London: Pan Books, 1964.

Studying:

Maddox, H. *How to Study*. London: Pan Books.

Writing:

Compton, H. *Conveying Ideas*. London: Cleaver-Hume, 1962.

Cooper, B. M. *Writing Technical Reports*. Baltimore: Penguin Books (Pelican), 1964.

Communication:

Aranguren, J. L. *Human Communications*. London: Weidenfeld & Nicolson, 1967.

Dean, H. H. & Bryson, K. D. *Effective Communication*. Englewood Cliffs, N. J.: Prentice-Hall, 1961.

Leyton, A. C. *The Art of Communication*. New York: Pitman, 1968.

Miller, G. A. *The Psychology of Communication*. Baltimore: Allen Lane, The Penguin Press, 1968.

Williams, R. *Communications*. Baltimore: Penguin Books, 1970.

NOTES FOR TEACHERS

This section is designed to provide teachers and college lecturers who intend to run their own courses and to use this book with some practical advice on the organization and instruction of reading efficiency training courses. This should enable them to see that there is no "magic" technique in teaching a reading improvement course and that the amount of preparation involved is no greater than is involved in devising any other course of instruction aimed at the improvement of skills. It should also enable them to avoid wasting time by using inappropriate and ineffective teaching methods.

Preparation

The essential information the instructor needs to give his students is contained in this book, but it may be useful to consult some of the books listed in the Further Reading section. On this basis, the tutor can build as his experience in running courses provides him with additional material.

It may be advisable to run one or two pilot courses with volunteers to gain familiarity with the procedures used in the course and to experience some of the problems readers may encounter on the course. Pilot courses also help in deciding whether the scheme of work suggested later in this section needs some modification to suit local conditions.

Essential Requirements

The essential requirements for a reading efficiency training course, run on the lines suggested here, are simply that every student should have a copy of this book, a notebook, and a pen or pencil.

Reading exercises can be timed in one of three ways:

a) by using a large timing device that records minutes and seconds, placed where every student can see it. By this method all students will have to begin reading at the same time.

b) by allowing the students to time themselves, if they each have stopwatches or wrist-watches with second hands. This method permits greater flexibility by allowing students to proceed at their own paces.

c) by using a blackboard "clock" marked as follows:

MINUTES	SECONDS			
	0	5	10	15
	20	25	30	35
	40	45	50	55

Using his own stop- or wrist-watch, the tutor indicates each 5-second interval and writes in the figure for each minute as it passes. The student then knows that if, when he finishes reading a passage, the tutor is pointing to 25 seconds and the figures "1 2 3" are written in the minutes space, he has taken 3 minutes and 25 seconds to read the passage. By this method, all the students must begin reading at the same time, but it has the advantage of being very inexpensive and yet effective.

Some other minor points are worth stating here, since they are often overlooked and can materially affect the atmosphere of a course. The room should be an appropriate size for the group. Rooms that are too large or too small hinder the development of a comfortable, informal atmosphere, which is important because of the mental demands made by the course. Results can also be affected if this point is overlooked. Desks or tables should be arranged in a U or O shape so that each student can see both the instructor and his fellow students. This is important for the discussions, which should be encouraged if time permits, on the instruction and the reading exercises. However, armchairs and similar kinds of furniture should be avoided because students may find it difficult to perform well on the reading exercises in surroundings that are too relaxed or informal. Pleasant, functional surroundings will give the best results.

Size of Groups

Groups of fewer than six students may limit discussion and groups that contain more than twenty students should be avoided if at all possible. Large groups usually progress less well than smaller ones, perhaps because of the reduced contact between the tutor and individual students. The best group size seems to be between twelve and fifteen students. All members of a group should be volunteers. All should currently have more reading matter to deal with than they can handle effectively. This will help to provide the kind of motivation essential for success. However, fairly successful courses have been held with skeptical and reluctant students.

Length of Course

The course in this book is designed to be spread over about six weekly meetings of two hours each (see A Suggested Scheme of Work below), though courses as short as four weekly meetings can be successful. Each teacher will be able to tailor the course to fit the time that he has available, but total teaching time should not be less than twelve hours. Longer courses have the advantage of providing more time for discussion.

In schools, it may be better to phase the instruction and the practice over one academic year, so that gradual progress is permitted and so that only a proportion of the weekly time available for English lessons need be occupied by the course.

In business firms, the course will probably have to be condensed into one-hour weekly meetings over five or six weeks. More time is desirable, but, if trainees are prepared to carry out most of the practice in their own time, the results from a short course should still make reading efficiency training a useful part of any line, general, or supervisory management education and training program.

Wherever courses are held, course meetings should be at weekly intervals and should be not less than thirty minutes nor more than two hours in length. Meetings shorter than one hour will mean that discussion will have to be curtailed. Results will be better if course meetings are held during the second half of the morning, though afternoon and even evening meetings can be successful. Meetings in the early evening, however, held soon after the participants have finished a day's work, should be avoided if at all possible.

Methods of Teaching

There is still, after more than fifty years of research and experiment in the running of reading efficiency courses, no evidence that courses using visual and mechanical aids in the form of pacers, films, tachistoscopes, or teaching machines are any more effective than those which do not. The only visual aid that the student needs is the progress graph and the only mechanical aid he requires is the stopwatch (or some other method of timing the reading exercises).

It is, in fact, important that the reading exercises should be carefully timed and tested and the results entered regularly on the Progress Graphs. Trainees can then see quickly how much they have progressed each week.

A simple combination of short periods of instruction, practice, and discussion provides a highly effective method of teaching reading efficiency. There is in the opinion of many tutors something peculiarly inappropriate about courses that purport to train reading skills by being machine- or film-centered rather than book-centered. No teacher should feel that he has to commit himself to the expenditure of large amounts of money in order to mount a reading improvement course. It is simply not necessary.

If, however, the finance is readily available for the purchase of equipment, no great harm will be done. Some items of equipment may even serve to introduce an element of novelty into a course and may help to encourage less highly motivated students. It should be remembered that the two most valuable elements in a course of this kind are the student's desire to improve and practice on a wide range of materials. Any means that will promote these elements are at least worth trying. Above all, the instructor should use those teaching methods and aids in which he himself has confidence, for there is some evidence that this confidence has a significant effect upon the success of a course.

Visual and Mechanical Aids

For those who do decide to purchase special equipment, some information about the main types currently available may be useful. There are four main kinds:

Films. Films usually project eight or nine lines of reading matter at a time on to a screen and segments of the material are successively revealed. Films are usually available in series and each film runs at a predetermined reading speed.

Teaching Machines. These combine the technique used in films and those of the tachistoscope. They have the advantage that progress is individual; with films, however, the whole group has to read at a certain speed.

Pacers. Pacers are devices for preventing regressions in eye movements and for urging the reader forward; other aids try to control the reader's eye movements much more strictly. There are two chief types of pacers. On one, reading matter printed on rolls is wound, at a pre-set speed, from one roller to another inside a plastic case. A glass panel enables a reader to see about a dozen lines at any one time. The reading matter passes the panel rather as the prologue on some old films used to move up the screen. On the other, which can be attached to a book, a bar moves down the page, again at a speed desired by the reader, thus preventing regressions.

Tachistoscopes. These instruments are sometimes called "flashers." They usually take the form of a flat plastic case with a small rectangular slot on the front near the bottom. A card is inserted in the case and a spring-loaded mechanism enables groups or numbers of words to be flashed before the reader's eyes at speeds up to 1/100th of a second as the card passes through the case. Tachistoscopes are held to increase a reader's speed and span of perception, but there is no evidence that the gains made are transferred to normal reading situations to any significant extent.

A Suggested Scheme of Work

To assist instructors in planning a course based on this book, a specimen scheme of work for a short course of six to ten weekly two-hour meetings may be helpful:

Week	*Content*
1.	Chapter 1
2.	Chapter 2
3.	Chapter 3
4.	Chapter 4
5.	Chapter 5
6.	Chapter 6
7.	Skimming—further discussion and practice
8.	Studying—further discussion and practice
9.	Critical Reading—further discussion and practice
10.	Three months' test.

In practice, there may be variations in progress with different groups. Chapters 3 and 5 in particular may stimulate discussion and if necessary they can be completed in weeks 6 and 10, respectively. The work set for these weeks will not usually take two hours, so the opportunity exists for further discussion of any problems, passages, or topics which students wish to return to. The more time that can be found for discussion, particularly after the first four weeks, the better; this helps to sustain interest in the course and to develop new reading interests in the students.

Specimen Time Allowance

Some idea of how the time in a typical two-hour session (except in weeks 1, 6, or 10) is allocated can be obtained from the following plan (timings are approximate, of course):

Time Taken		*Item*
5 minutes	1.	Registration, etc.
10 minutes	2.	Revision of previous week and first part of instruction.
20 minutes	3.	First reading exercise and discussion.
10 minutes	4.	Second part of instruction.
20 minutes	5.	Second reading exercise and discussion.
10 minutes	6.	Third part of instruction.
20 minutes	7.	Third reading exercise and discussion.
5 minutes	8.	Revision of week's instruction.
15 minutes	9.	Discussion of points raised by students.
5 minutes	10.	Practice work for coming week.
120 minutes		

Records of Practice

Students should be encouraged to keep records of the practice reading done in their own time during a course. There are two reasons for this. First, it reminds them of the importance of practice, and second, it enables the instructor to introduce an element of individual learning, especially if the kind of Record of Practice form illustrated opposite is used. The forms can be completed during the class meeting, handed in to the teacher at the end, and returned with comments the following weeks. In this way the essentially confidential nature of the course is also preserved.

This method will enable the instructor to gain greater experience of the problems individual students encounter in the process of improving their reading skills. Since each group of students is unique in that no two groups encounter exactly the same difficulties, this information will be invaluable in helping the teacher to tailor the course more precisely to the needs of individual students.

RECORD OF PRACTICE

Name of Student: ....

Group: ....

1. <u>Practice</u>: State briefly
   a) the material you have used for practice: ....
   ....
   b) how long you have practiced each day ....
   c) any problems you have encountered in your practice: ....
   ....

2. <u>Today's Results in Class (Average)</u>

   Reading speed: .... words per minute

   Comprehension: .... out of ten.

3. Instructor's Comments: ....
   ....
   ....

Integration with Other Studies

Where possible, a reading efficiency training course should be integrated into a general program of communication skills improvement. There is evidence that suggests that reading and listening skills are closely connected and that when one is improved the other also benefits. It is probably the case that all communication skills—thinking, reading, writing, listening, and speaking—are related and that, if reading improvement is set into a wider context, the benefits for the individual will be considerable.

Continuation and Follow-Up Work

This is particularly important, and teachers should try to ensure that students complete the continuation work set in this book. Improvements can disappear unless some attempt is made to remind students of the need to make their improved skills work for them. Continuation work ensures that the improvements gained are maintained or even further improved.

Index